# Western Fertilizer Handbook

## Eighth Edition

# Western Fertilizer Handbook

Produced by

SOIL IMPROVEMENT COMMITTEE
CALIFORNIA FERTILIZER ASSOCIATION

Interstate Publishers, Inc.

**WESTERN FERTILIZER HANDBOOK,** Eighth Edition. Copyright © 1995 by the California Fertilizer Association, 1700 I Street, Suite 130, Sacramento, CA 95814. All rights reserved. Prior editions: 1953, 1954, 1961, 1965, 1975, 1980, 1985. Printed in the United States of America.

*Library of Congress Catalog Card No. 94-77467*

ISBN 0-8134-2972-2

1  2  3

4  5  6

7  8  9

Order from

INTERSTATE PUBLISHERS, INC.

510  North Vermilion Street
P.O. Box 50
Danville, IL 61834-0050

Phone: (800) 843-4774
Fax: (217) 446-9706

# Foreword

The California Fertilizer Association was established in 1923 with the purpose of promoting progress within the fertilizer industry in the interest of an efficient and profitable agricultural community. This continues to be its purpose today.

Activities of the association include developing and disseminating new information to its members and others; supporting production-oriented research programs to identify best management practices for farmers; promoting the teaching of agronomic topics at our schools, colleges, and universities; and maintaining effective communication among industry, university, and other state and federal agencies.

Many of the above activities are carried out by the Soil Improvement Committee, which was established by the membership in the mid-1920's. The Soil Improvement Committee produced the first edition of the *Western Fertilizer Handbook* in 1953. With later editions, total distribution now exceeds 175,000 copies.

There are other associations and institutes that sponsor programs similar to those of the California Fertilizer Association. These include:

1. Agricultural Retailers Association, 339 Consort Drive, Manchester, MO 63011.

2. Arizona Agricultural Chemicals Association, 2406 South 24th Street, Suite E-103, Phoenix, AZ 85034.

3. Canadian Fertilizer Institute, 360 Albert Street, Suite 1540, Ottawa, Ontario, Canada K1R 7X7.

4. Foundation for Agronomic Research, 655 Engineering Drive, Suite 110, Norcross, GA 30092-2821.

5. Idaho Fertilizer and Chemical Association, P. O. Box 190130, Boise, ID 83719-0130.

6. The Fertilizer Institute, 501-Second Street, N.E., Washington, DC 20002.

7. National Agricultural Chemicals Association, 1155 15th Street, N.W., Washington, DC.

8. Potash & Phosphate Institute, 655 Engineering Drive, Suite 110, Norcross, GA 30092-2821.

9. Rocky Mountain Plant Food and Agricultural Chemical Association, 2150 S. Bellaire St., Denver, CO 80222.

10. The Sulphur Institute, 1140 Connecticut Avenue, N.W. Suite 612, Washington, DC 20036.

11. Western Agricultural Chemicals Association, 3835 N. Freeway Blvd., Suite 140, Sacramento, CA 95834.

12. Western Canada Fertilizer Association, 9250-120 Street, Surry, British Columbia, Canada V3V 4B7.

13. Western Fertilizer & Chemical Dealers Association, Box 1301, Brandon, Manitoba, Canada R7A 6N2.

# Acknowledgements

The eighth edition of the Western Fertilizer Handbook carries on the tradition of its predecessors in presenting information directly applicable to the agricultural industry. The book presents information on fertilization, nutrient management, and related topics based on the fundamentals of biological and physical sciences.

The eighth edition differs significantly from previous editions in two areas. First, much of the horticultural information has been deleted because a horticulture edition of the *Western Fertilizer Handbook,* specifically addressing this sector of our industry, was printed in 1990. Second, there is considerably more focus on the relationships of fertilizer application and crop management to environmental quality.

Much of the information contained in the eighth edition of the *Western Fertilizer Handbook* was taken directly from or updated from previous editions. The Editorial Committee, therefore, gratefully acknowledges the original contributions of the many excellent scientists to these earlier editions. Also, other members of the California Fertilizer Association and colleagues of the association contributed many ideas and photos. They are also gratefully acknowledged.

EDITORIAL COMMITTEE

Albert E. Ludwick, Chairman
Larry C. Bonczkowski
Carl A. Bruice
Keith B. Campbell
Robert M. Millaway
Steven E. Petrie
Irvine L. Phillips
J. Julian Smith

# Table of Contents

# Introduction

Western agriculture may be described in many ways—dynamic, aggressive, modern, changing, diverse. . . . Certainly, its trademark is the production of quality, high-yielding crops. Tree fruit and nut crops, along with vine, forage, grain, vegetable, and fiber crops, are extensively grown in the West, and many are distributed to consumers worldwide. The fertilizer industry contributes significantly to this production and has for many years. It is proud of its role and looks forward to continuing success.

The *Western Fertilizer Handbook* is written specifically for western agriculture and for that individual seeking practical, production-oriented, problem solving information. Its language is concise and straightforward. Basic information is included on properties of soil, water, and plants. Fertilizer products, their properties, and their management are presented, as is related information on soil amendments and irrigation water quality. All of this is framed in the context of western agriculture, that is, primarily high-yielding, irrigated crops grown on soils frequently having a pH above 7.0.

The *Western Fertilizer Handbook* is not a recipe book in which information is applied directly to a production situation without further consideration. Rather it is intended as an educational text to develop a better understanding of agronomic principles and practices, and to provide a convenient reference on soil, water, plants, and fertilizer. Guides within chapters, such as those on soil and tissue testing, may or may not be the most appropriate to use in a specific situation.

## FOOD FOR A GROWING POPULATION

There is an urgency to produce higher yields in the future. This may seem inconsistent for a nation that has never experienced a serious food shortage and periodically is faced with surpluses. These surpluses are, however, an artifact of world economic problems, marketing difficulties, logistics of distribution, and politics.

According to the Food and Agriculture Organization (FAO), there are about 450 million undernourished people in the world today. If the current trend continues, there will be 600 million undernourished people by the year 2000.

A film entitled *A Gift of Harvest*[1] states that, should world population double in the next 35 years, as many experts predict, we will have to produce as much food in the next three decades as has ever been produced. Even if this estimate is substantially off, the challenge facing those of us involved in food production is almost incomprehensible. All indications are that world agricultural production will have to double in less than 20 years to meet expanding world food needs.

How can we do this? The FAO estimates that about one-fourth of the added requirements will come from new land being brought into production for the first time. The remaining three-fourths of the new food requirement will have to be met by increasing yields on currently cropped land.

Additional fertilization alone does not guarantee higher yields. Rather, the careful integration of a balanced and sound fertilizer program with all the other production inputs will result in high yield. Such an approach of balanced inputs and management is frequently termed "Best Management Practices." It is important that each input be consistent with the yield goal. In this way, all inputs, that is, fertilizer, irrigation water, pesticides, etc., will be efficiently used. Frequently, this leads to positive interactions among inputs. Interactions occur when crop response to two or more inputs used together is greater than the sum of their individual responses. It also leads to more of the inputs being utilized by the growing crop and less remaining following harvest that might pose an environmental hazard.

---

[1]Produced by the National Agricultural Chemicals Association. See listing at end of Forward.

# MAXIMUM ECONOMIC YIELDS

Maximum economic yields (top profit yields) come from growers using Best Management Practices. They are very important to farmers who want to make a reasonable profit. As business people, farmers are in a unique position. They operate on a narrow profit margin—frequently too narrow for their own well-being. Of the three basic factors determining profitability—yield level, selling price, and production cost—farmers have relatively little control over the latter two. The opportunity for greater profits lies mostly with yield level. Higher yields increase production efficiency. This is because in the range most farmers are operating, over 80 percent of the costs of production are about the same regardless of yield. More production per acre means less cost per unit of output, that is, lower cost to produce each pound of grain, each bale of cotton, each ton of alfalfa, etc.

Producing maximum economic yields translates into more food for a growing world population and less expensive food. The United States can boast some of the world's best farmers. We spend only 10 percent of our disposable income on food, compared to 18 percent in Japan, 32 percent in Mexico, 48 percent in China, and 53 percent in India according to the USDA.

Fertilizer is a key ingredient of high yields. Historically, it has given farmers a high return on their investment. It still does today. Experts estimate that 30 to 50 percent of today's crop production comes directly from fertilizer. Yet for many irrigated crops of the West, it amounts to only 10 or 15 percent of total production costs. As yields continue to increase, fertilizer will become even more important because the gap between native soil fertility and crop nutrient requirements will widen. Also, each additional year of cropping depletes those nutrients not replaced through fertilization.

In striving for higher, more efficient yields, an old saying sums up the role of fertilizer very well: "A fertile soil is not always a productive soil, but a productive soil is always a fertile soil."

# Chapter 1

# Soil – A Medium for Plant Growth

Food comes from the earth. The land with its waters gives us nourishment. The earth rewards richly the knowing and diligent, but punishes inexorably the ignorant and slothful. This partnership of land and grower is the rock foundation of our complex social structure.

W. C. Lowdermilk

## WHAT IS SOIL?

As a medium for plant growth, soil can be described as a complex natural material derived from disintegrated and decomposed rocks and organic materials that provides nutrients, moisture, and anchorage for land plants.

The four principal components of soil are minerals, organic matter, water, and air. These are combined in widely varying amounts in different kinds of soil, and at different moisture levels. A representative western soil, at an ideal moisture content for plant growth, is nearly equally divided between solid materials and pore space, on a volume basis. The pore space contains nearly equal amounts of water and air. Figure 1-1 shows a schematic representation of such a soil.

1

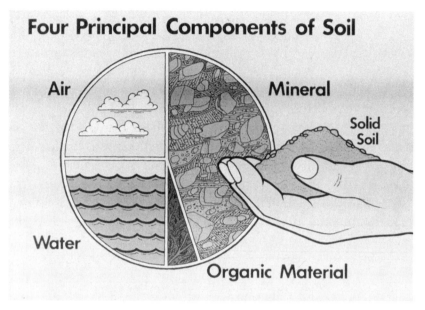

Figure 1-1. Volumetric content of four principal soil components for a representative western soil at an ideal moisture content for plant growth.

## *HOW SOILS ARE FORMED*

The development of soils from original rock materials is a long-term process involving both physical and chemical weathering, along with biological activity. Hundreds or thousands of years are required for the formation of just one inch of topsoil. The widely variable characteristics of soils are due to differential influences of the soil formation factors:

1. Parent material – material from which soils were formed.

2. Climate – temperature and moisture.

3. Living organisms – microscopic and macroscopic plants and animals.

4. Topography – shape and position of land surfaces.

5. Time – period during which parent materials have been subjected to soil formation.

The initial action on the original rock is largely mechanical – cracking and chipping due to temperature changes and abrasion by wind and water. As the rock is broken into smaller particles, the total surface area exposed to the atmosphere increases. Chemical action of water, oxygen, carbon dioxide, and various acids further reduces the size of rock fragments and changes the chemical composition of many of the resulting particulate materials. Finally, the action of microorganisms and high plant and animal life contributes organic matter to the weathered rock material, and a true soil begins to form.

Since all of these soil-forming agents are in operation constantly, the process of soil formation is continual. Evidence indicates that the soils upon which we depend to produce our crops required hundreds or thousands of years to form. In this regard, we should consider soil a nonrenewable resource, measured in terms of our life span. Thus, it is very important that we protect our soils from destructive, erosive forces and nutrient depletion, which can rapidly destroy the product of hundreds of years of nature's work.

## SOIL PROFILE

A vertical section through a soil typically presents a layered pattern. This section is called a *profile,* and the individual layers are referred to as *horizons.* A representative soil has three general horizons, which may or may not be clearly discernible (Figure 1-2).

The uppermost horizon includes the *surface soil,* or *topsoil,* and is designated the *A* horizon. The next successive horizon, underlying the surface soil, is called the *subsoil,* or *B* horizon. The combined A and B horizons are referred to as the *solum.* Underlying the B horizon is the *parent material,* or *C* horizon.

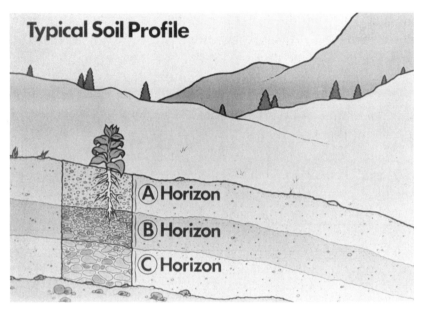

Figure 1-2. This typical soil profile illustrates the three general soil horizons.

These three horizons, together with the unweathered, unconsolidated rock fragments lying on top of the bedrock, are called the *regolith*.

Soil profiles vary greatly in depth or thickness, from a fraction of an inch to many feet. Normally, however, a soil profile will extend to a depth of about 3 to 6 feet. Other soil characteristics, such as color, texture, structure, and chemical composition also exhibit wide variations among the many soil types.

The surface soil (A horizon) is the layer most subject to climatic and biological influences. Most of the organic matter accumulates in this layer, which usually gives it a darker color than the underlying horizons. Commonly, this layer is characterized by a loss of soluble colloidal materials, which are moved into the lower horizons by infiltrating water, a process called *eluviation*.

The subsoil (B horizon) is a layer which commonly accumulates many of the materials leached and transported from the surface soil. This accumulation is termed *illuviation*. The deposition of materials such as clay particles, iron, aluminum, calcium carbonate, calcium sulfate, and other salts creates a layer that normally has more compact structure than the surface soil. This often leads to restricted movement of moisture and air within this layer, which can limit root growth.

The parent material (C horizon) is the least affected by physical, chemical, and biological agents. It is very similar in chemical composition to the original material from which the A and B horizons were formed. Parent material that has formed in its original position by weathering of bedrock is termed *sedentary* or *residual*. That which has been moved to a new location by natural forces is called *transported*. This latter type is further characterized on the basis of the kind of natural force responsible for its transportation and deposition. When water is the transporting agent, the parent materials are referred to as *alluvial* (stream-deposited), *marine* (sea-deposited), or *lacustrine* (lake-deposited). Wind-deposited materials are called *aeolian*. Materials transported by gravity are called *colluvial,* a category that is relatively unimportant with respect to agricultural soils.

Because of the strong influence of climate on soil profile development, certain general characteristics of soils formed in areas of different climatic patterns can be described. For example, much of the western United States has an arid or semi-arid climate, which results in the development of coarser-textured soils (more sand particles) than most of those developed in more humid climates. Also, many western soil profiles are less developed, since the amount of water percolating through the soil is generally much less than in more humid climates. Because of this, many western soils contain more calcium, potassium, phosphorus, and other nutrient elements than do more extensively weathered eastern soils.

Thus, the soil profile is an important consideration in terms of plant growth. The depth of the soil, its texture and structure, and its chemical nature determine to a large extent the value of the soil as a medium for plant growth.

# SOIL TEXTURE

Soils are composed of particles with an infinite variety of sizes and shapes. Individual mineral particles are divided into three categories on the basis of their size — sand, silt, and clay. Such a division is very meaningful, not only in terms of a classification system, but also in relation to plant growth. Many of the important chemical and physical reactions are associated with the surface of the particles. Surface area is enlarged greatly as particle size diminishes, which means that the smallest particles (clay) are the most important with respect to these reactions.

Two systems of classification of the various particle sizes (soil separates) are used. One is the U.S. Department of Agriculture system, and the other is the International system. The size ranges of the various soil separates for the two systems are shown in Table 1-1.

Table 1–1
Size Limits of Soil Separates[1]

| U.S. Department of Agriculture | | International | |
|---|---|---|---|
| Name of Separate | Diameter (Range) | Fraction | Diameter (Range) |
| | *(mm)* | | *(mm)* |
| Very coarse sand | 2.0 – 1.0 | | |
| Coarse sand | 1.0 – 0.5 | I | 2.0 – 0.2 |
| Medium sand | 0.5 – 0.25 | | |
| Fine sand | 0.25 – 0.10 | II | 0.20 – 0.02 |
| Very fine sand | 0.10 – 0.05 | | |
| Silt | 0.05 – 0.002 | III | 0.02 – 0.002 |
| Clay | Below 0.002 | IV | Below 0.002 |

[1]Source: USDA *Soil Survey Manual,* Agricultural Handbook No. 18.

Soil texture is determined by the relative proportion of sand, silt, and clay found in the soil. Twelve basic soil textural classes are recognized. A classification chart based on the percentages of sand, silt, and clay appears in Figure 1-3.

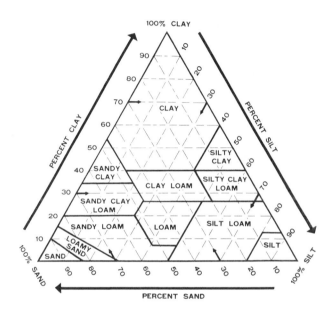

Figure 1-3. Soil textural triangle showing the percentages of clay (below 0.002 mm), silt (0.002 to 0.05 mm), and sand (0.05 to 2.0 mm) in the basic textural classes.

A textural class description of soils reveals much about soil-plant interactions, since the physical properties of soils are determined largely by the texture. In mineral soils, the exchange capacity (ability to hold plant nutrients) is closely related to the amount and kind of clay in the soil. The water-holding capacity is determined in large measure by the particle size distribution. Fine-textured soils (high percentage of silt and clay) hold more water than coarse-textured soils (sandy). Finer-textured soils often are more compact, have slower movement of water and air, and can be more difficult to work.

Medium-textured soils, such as loams, sandy loams, and silt loams, are probably the most suitable for plant growth. Nevertheless, there is only a general relationship between soil textural class and soil productivity, since texture is only one of the many factors that influence crop production.

## SOIL STRUCTURE

Except for sand, soil particles normally do not exist singularly in the soil, but rather are arranged into aggregates or groups of particles. The way in which particles are grouped together is termed *soil structure*.

There are four primary types of structure, based upon shape and arrangement of the aggregates (Figure 1-4). Where the particles are arranged around a horizontal plane, the structure is called *plate-like* or *platy*. This type of structure can occur in any part of the profile. Puddling or ponding of soils often gives this type of structure on the soil surface. When particles are arranged around a vertical line and bounded by relatively flat vertical surfaces, the structure is referred to as *prism-like* (prismatic or columnar). Prism-like structure is usually found in subsoils and is common in arid and semi-arid regions. The third type of structure is referred to as *block-like* (angular blocky or subangular blocky) and is characterized by approximately equal length in all three dimensions. This arrangement of aggregates is also most common in subsoils, particularly those in humid regions. The fourth structural arrangement is called *spheroidal* (granular or crumb) and includes all rounded aggregates. Granular and crumb structures are characteristic of many surface soils, particularly where the organic matter content is high. Soil management practices can have an important influence on this type of structure.

Soil aggregates are formed both by physical forces and by binding agents—principally products of decomposition of organic matter. The latter types are more stable and resist to a

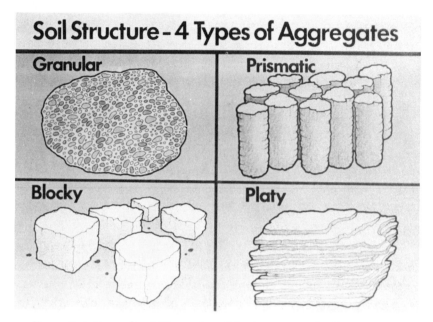

Figure 1-4. Generalized illustration of the four types of soil aggregates.

greater degree the destructive forces of water and cultivation. Aggregates formed by physical forces such as drying, freezing and thawing, and tillage operations are relatively unstable and are subject to quicker destruction.

Soil structure has an important influence on plant growth, primarily as it affects moisture relationships, aeration, heat transfer, and mechanical impedance of root growth. For example, the importance of good seedbed preparation is related to moisture and heat transfer — both of which are important in seed germination. A fine, granular structure is ideal in this respect.

The movement of moisture and air through the soil is dependent upon porosity, which is markedly influenced by soil structure. Granular structure provides adequate porosity for good infiltration of water and air exchange between the soil and the atmosphere. This creates an ideal physical medium for plant growth. However, where surface crusting occurs, or sub-surface claypans or hardpans exist, plant growth is hindered because of restricted porosity. Good management practices can improve soil structure and thereby create a better condition for crop production.

# SOIL REACTION (pH)

Soil reaction (acid, neutral, alkaline) refers to the relative concentration of hydrogen ions ($H^+$) and hydroxyl ions ($OH^-$) in the soil solution. An acid soil has a higher concentration of hydrogen ions than hydroxyl ions while an alkaline soil has more hydroxyl ions than hydrogen ions. A neutral soil has equal concentrations of both hydrogen ions and hydroxyl ions.

The actual concentrations of hydrogen or hydroxyl ions are extremely low. A soil with a pH of 7.0 has only 0.0000001 moles of hydrogen ions per liter of soil solution. This is a cumbersome way of reporting soil acidity and the term pH is used instead. The pH scale ranges from 0 to 14, with pH 7 being neutral; pH values below 7 are acid and pH values above 7 are alkaline. The lower the pH value, the more acid the soil and, conversely, the higher the pH above 7, the more alkaline the soil. The pH scale is logarithmic so each pH unit change represents a tenfold change in acidity or alkalinity. For example, a soil with a pH of 6 is ten times as acid as one with a pH of 7. A soil with a pH of 8 is ten times as alkaline as one with a pH of 7 and 100 times as alkaline as one with a pH of 6.

> Note: The term alkaline refers to soil pH above 7.0; the term alkali is an older term for soils with excessive sodium. See Chapter 9.

The soil reaction is important because it influences (1) nutrient availability, (2) solubility of toxic ions, and (3) microbial activity.

Nutrient availability varies at different pH values (Fig-

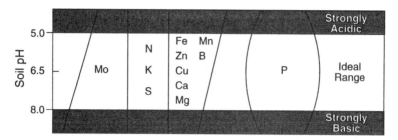

Figure 1-5. Relative mineral nutrient availability as affected by soil pH.

ure 1-5). The maximum availability of the primary nutrients (N-P-K) is greatest at a pH value between 6.5 and 7.5. The solubility of some phytotoxic elements, such as aluminum, increases at low pH values (below 5.5) which can reduce crop yields. Near neutral pH values favor the activity of many microorganisms responsible for essential soil biological activity.

There are four sources of soil acidity. First, basic cations such as calcium, magnesium, and potassium are leached from the top soil into the subsoil as a result of rainfall. Second, these cations are taken up and removed by crops. Third, the use of nitrogen fertilizers containing ammonium will lower the soil pH. Finally, the use of elemental sulfur for soil reclamation or plant nutrition will lower soil pH.

The calcium carbonate (free lime) found in many western soils acts as a buffer against the development of acid soils. This occurs because the solubility of calcium carbonate is increased as the acidity increases. This, in turn, raises the level of exchangeable calcium in the soil. This elevated soluble calcium replaces hydrogen ions which then combine with oxygen from the carbonate to form water and carbon dioxide. The buffering capacity of many western soils, combined with the low rainfall, explains why many western soils are alkaline.

Calcium carbonate can be used to raise the pH of acid soils through the same mechanism. See Chapter 9 for more information on correcting soil problems with amendments.

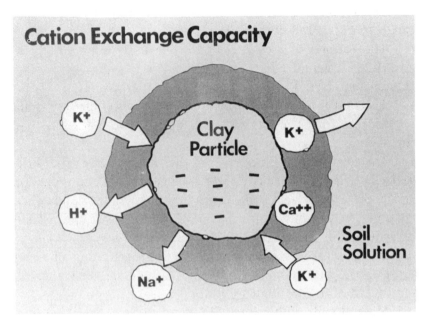

Figure 1-6. Schematic illustration of the exchange of cations between the negatively charged clay particles and the soil solution.

# CATION EXCHANGE CAPACITY

The cation exchange capacity (CEC) is an important measure of the fertility and potential productivity of a soil.

Due to their chemical structure, clay particles and soil organic matter have a net negative charge. This means that cations (positively charged ions) can be attracted to and held on the surface of these soil materials. Cations in the soil solution are in dynamic equilibrium with the cations adsorbed on the surface of the clay and organic matter. This exchange is shown in Figure 1-6. The CEC is a measure of the quantity of cations that can be adsorbed or held by a soil.

Soils contain varying amounts and different kinds of clay and organic matter, so the total CEC is widely variable among different soil types. Soil organic matter has a high CEC and soils with high organic matter content typically have a higher CEC than soils low in organic matter.

The cations of greatest significance with respect to plant growth are calcium ($Ca^{++}$), magnesium ($Mg^{++}$), potassium ($K^+$), ammonium ($NH_4^+$), sodium ($Na^+$), and hydrogen ($H^+$).

The first four cations are plant nutrients and are important for plant growth. The last two have a pronounced effect on soil chemical and physical characteristics.

The relative amount of each of the cations adsorbed on the surface of clay particles is closely associated with important soil properties. Highly acid soils have a high percentage of adsorbed hydrogen while soils with a favorable pH (6 to 8) have a high percentage of adsorbed calcium ions. Soils that are high in sodium ions are dispersed and resist the infiltration of water while those with a high percentage of calcium ions are well aggregated and have high infiltration rates.

Mineral soils with a high CEC tend to be more fertile than those with a lower CEC. The nutrients are less likely to be lost due to leaching and the soil has a greater ability to store and supply nutrients to crops. The approximate CEC range for some typical soil texture classes is shown in Table 1-2. The range of values within a textural class is due, in part, to differences in organic matter content.

### Table 1–2
### Typical CEC of Some Soil Texture Classes

| Soil Texture | Typical CEC Range |
|---|---|
|  | *meq/100 gms* |
| Sand | 2 – 6 |
| Sandy Loam | 3 – 8 |
| Loam | 7 – 15 |
| Silt Loam | 10 – 18 |
| Clay and Clay Loam | 15 – 30 |

# SOIL ORGANIC MATTER

Soil organic matter is one of the major components of soil productivity and soil fertility; its importance to crop production cannot be overstated. Soil organic matter:

1. Helps strengthen soil aggregates, thus improving soil tilth and structure.

2. Improves aeration and water infiltration.

3. Increases water-holding capacity.

4. Provides significant amounts of cation exchange capacity.

5. Provides buffering against rapid changes in soil reaction when acid or alkaline forming materials are added to soil.

6. Forms stable organic compounds which can increase the availability of micronutrients.

7. Provides a source of many plant nutrients.

8. Provides a food source for soil microorganisms.

Western soils typically contain lower amounts of organic matter than soils formed in higher rainfall areas. Many western soils have only 1 to 2 percent organic matter while soils in the eastern United States may have as much as 7 to 10 percent organic matter. Some western soils, however, such as river delta soils or soils in old lake beds, may have as much as 50 to 75 percent organic matter.

Soil organic matter consists of plant and animal residues in various stages of decomposition, living organisms, and the substances synthesized by those organisms. The composition of soil organic matter may be broken into three major categories: polysaccharides, lignins, and proteins (Table 1–3). The polysaccharides include cellulose, hemicellulose, sugars, starches, and pectic substances. Lignins are complex materials derived from woody tissues of plants. Proteins are the principal nitrogen-con-

Table 1–3
Common Constituents of Soil Organic Matter
and Relative Rates of Decomposition

| Organic Constituents | Approximate Percent of Total Organic Matter | |
| --- | --- | --- |
| | | Rapidly decomposed |
| Sugars, starches, simple proteins | 1–5 | ↑ |
| Crude proteins | 5–20 | |
| Hemicelluloses | 10–25 | |
| Cellulose | 30–50 | ↓ |
| Lignins, fats, waxes | 10–30 | Very slowly decomposed |

taining constituents of organic matter. These three classes of materials are the sources of food for soil microorganisms.

Organic residue is decomposed in the soil by living organisms, primarily bacteria and fungi. Each group of organisms becomes a dominant factor as decomposition proceeds. These microorganisms excrete various compounds which can bind soil particles together. Larger organisms such as earthworms and insects ingest organic residue and soil particles, thereby binding them together into stable aggregates.

Soil organic matter is a source of nitrogen, phosphorus, and sulfur for plant nutrition. For these nutrients to become available to plants the soil organic matter must be mineralized; mineralization converts the nutrients in organic matter into the inorganic ions taken up by plants.

Mineralization is carried out by soil microorganisms in the generalized equation shown below.

(soil organisms)

$$\text{organic matter} + H_2O + O_2 \rightarrow CO_2 + H_2O + NH_4^+$$
$$+ H_3PO_4 + H_2S$$
$$+ \text{other gases} + \text{energy}$$

As a rule of thumb, decomposition of organic matter will release about forty pounds of nitrogen for each one percent organic matter in the soil. In addition to the nitrogen released, the decomposition of organic matter will release small amounts of phosphorus and sulfur.

The soil organisms responsible for decomposing soil organic matter require nutrients, especially nitrogen. The carbon:nitrogen ratio of soil organic matter is usually taken to be 10:1. The process of decomposing organic material with a high carbon:nitrogen ratio will tie-up nitrogen and will reduce the availability of nitrogen to plants. This process is called immobilization. Many crop residues commonly returned to the soil have high carbon:nitrogen ratios and will immobilize soil nitrogen unless supplemental nitrogen is applied. When corn stover, grain straw, cotton trash, or other non-leguminous residues are incorporated into soil, a rule of thumb is to apply 15 to 20 pounds of nitrogen per ton of residue.

# SOURCES OF ORGANIC MATTER

The organic matter found in most agricultural soils is derived from crop residues and manures. Residues with lowest carbon:nitrogen ratios decompose the most rapidly. These come from cover crops (green manures) such as legumes, relatively young grasses grown as a cover crop, and mustards. These crops usually provide nitrogen in excess of microbial needs and can supply crops with readily available nutrients. Animal manures can also be used as sources of soil organic matter, but care must be taken because manures may contain high levels of salts which can have deleterious effects on soil conditions and crop growth. Manure should always be analyzed before being used as a fertilizer.

# ORGANIC SOILS

Organic soils are those soils which contain more than 20 to 30 percent organic matter, depending upon the texture of the mineral fraction of the soil. Organic soils develop when conditions are such that normal decomposition of soil organic matter is extremely slow. These conditions occur in shallow lakes, swampy areas, and peat bogs. Organic soils appear almost black when moist. They are quite lightweight when dry with bulk density often less than 0.5 gm/cc compared to mineral soils with bulk density of 1.2 to 1.4 gm/cc. Organic soils can hold large amounts of water; up to four times their weight in water.

Organic soils can be quite productive if they are managed properly. Judicious use of fertilizer is usually necessary to achieve optimum production. Nitrogen fertilization may not be needed initially, but after a few years of cropping, the use of nitrogen fertilizer is usually necessary for good yields. Phosphorus and potassium are usually low in organic soils and must be applied for high yields. Deficiencies of micronutrients such as zinc, copper, and manganese are quite common on organic soils.

# SOIL MICROORGANISMS

In addition to their role in soil-forming processes, soil organisms make an important contribution to plant growth through their effects on the nutrient cycling in soil. Particularly important in this respect are the microscopic plants (microflora) which decompose organic residue releasing nutrients for growing plants.

Some important kinds of microorganisms are bacteria, fungi, and algae. All of these are present in the soil in very large numbers when conditions are favorable. A gram of soil (about 1 cubic centimeter) may contain as many as 4 billion bacteria, 1 million fungi, and 300,000 algae. These microorganisms are important in the decomposition of organic materials, the subsequent release of nutrients, and fixation of atmospheric nitrogen.

Soil bacteria are of special interest because of their many varied activities. A major group of bacteria function in decomposing organic materials (heterotrophic bacteria). There is also a smaller group which obtain their energy from the oxidation of materials such as ammonium, sulfur, and iron (autotrophic bacteria). The latter group is responsible for nitrification (the conversion of ammonium to nitrate) in the soil, a process which is vital in providing nitrogen for the growth of plants.

Nitrogen-fixing bacteria also play an important role in the growth of higher plants. They are capable of converting atmospheric nitrogen into reduced nitrogen, which is available to the plant in which the bacteria is living. Nodule bacteria (*Rhizobia* spp.) live in the roots of legumes, deriving their energy from the host plant, and fixing nitrogen from the soil atmosphere. The amount of nitrogen that is fixed depends upon many factors, such as the legume grown, the strain of bacteria infecting the roots, the overall plant health, and growing conditions. Values as high as 250 to 300 pounds per acre have been reported, but the average is much less: perhaps 80 to 100 pounds of nitrogen per acre. There are free-living bacteria (*Azotobacter* and *Clostridium*) which also fix nitrogen, but the amount fixed is much less.

## *SOIL MANAGEMENT*

Our soil resource is one of the great natural treasures of the United States. Manage is defined as "to handle with a degree of skill" or "to treat with care"; good soil management must surely encompass these definitions. These definitions imply using the best available knowledge, techniques, materials, and equipment for growing plants. Through wise management, we can produce abundant crops and ornamental plants, improve the soils and enhance the environment, thus providing a priceless legacy for future generations.

Tillage serves many important purposes. Seedbed preparation, weed control, incorporation of plant residues and fertilizers, breaking soil crusts and hardpans, and improving soil characteristics for irrigation and erosion control are some of the major reasons for tillage. Inappropriate tillage can cause soil compaction, increase erosion, and reduce soil quality and productivity.

Soil conservation is another important management practice that deserves careful attention. It has been estimated that annual soil erosion losses in the United States approach 4 billion tons of topsoil. This is equivalent to the total loss of topsoil six inches deep from 4 million acres. Soil management practices such as contour farming, strip planting, windbreaks, cover cropping, reduced tillage, terracing, and improved plant residue management are some of the important soil management practices that should be utilized where appropriate.

Figure 1-7. Deep tillage is an important management practice on many western soils to improve water penetration.

Proper residue utilization is a key to good soil management. Plant residues returned to the soil help maintain soil productivity through maintenance of soil organic matter. Plant residues are one of the most valuable tools growers can use to minimize wind and water erosion.

Soil management is one component of a total crop management system. Best Management Practices (BMP's) are those crop, fertilizer, pesticide, and water management practices which lead to increased fertilizer efficiency, minimize the loss of inputs, and maintain or increase crop yields. The adoption and utilization of BMP's will result in both the greatest economic return to growers and the maximum protection of the environment.

## SUPPLEMENTARY READING

1. *Fundamentals of Soil Science,* Seventh Edition. H. D. Foth. John Wiley & Sons, Inc. 1984.

2. *The Nature and Properties of Soils,* Eighth Edition. N. C. Brady. The Macmillan Company. 1974.

3. *Soil Conservation,* Second Edition. N. Hudson. Cornell University Press. 1981.

4. *Soil Erosion and Crop Productivity.* R. F. Follett and B. A. Stewart, ed. American Soc. of Agronomy. 1985.

5. *Soil Genesis and Classification, Second Edition. S. W. Buol, F. D. Hole and R. J. McCracken. Iowa State University Press. 1980.*

# Chapter 2

# *Water and Plant Growth*

All water used for irrigation contains some dissolved salts. The suitability of water for irrigation generally depends on the kinds and amounts of salts present. All salts in irrigation waters have an effect on plant-soil-water relations, on the properties of soils, and indirectly on the production of plants.

A user of irrigation water should know the effects that water quality and irrigation practices have on:

1. Salt content (salinity) of the soil.

2. Sodium status (sodicity) of the soil.

3. Rate of water infiltration.

4. Toxic element status.

5. Soil nitrate status and groundwater quality.

6. Fouling, or plugging, of microirrigation systems, where used.

It is difficult to isolate these factors one from another because some of them are interrelated. Furthermore, certain salts may have an effect on more than one of the factors listed above.

## *IRRIGATION*

Irrigation water is applied to soil to replenish water removed from the soil by evaporation, by growing plants, and by drainage below the root zone. It is applied in a number of ways. The method of application depends primarily upon the plants to be grown, the depth and texture of the soil, the topography of the land, and the cost of the water. The amount of water used and how often it is applied are determined by the method of irrigation employed, plant needs, leaching requirements, the soil's innate water holding capacity, and climatic considerations. Therefore, successful irrigation requires careful management of plants, soils, and water.

Figure 2-1. Irrigation water is applied to replace the water removed from the soil.

## *SOIL WATER BEHAVIOR*

In a well drained soil, most water is held as a film around each soil particle. The thinner these films are, the more tightly the water is held and the greater the suction needed to remove the water.

Immediately following an irrigation, the films of water are thick and, therefore, are not tightly held by the soil. The amount of water held is called the *saturation percentage.*

In about two or three days, with free drainage, about one-half of this weakly held water moves deeper into the soil, and free drainage ceases. The soil water content at this point is called *field capacity.* The films of water are now thinner and are held more tightly.

·     Below field capacity, gravity is no longer a significant force in moving water in the soil. Most of the water is then removed by the roots of growing plants. Plants will remove about one-half of the water held at field capacity. At that point the soil holds water so tightly that plants cannot extract it, thus causing them to wilt. That point is called the *permanent wilting point.*

The water content of a soil saturated with water, its saturation percentage, is about twice the field capacity and about

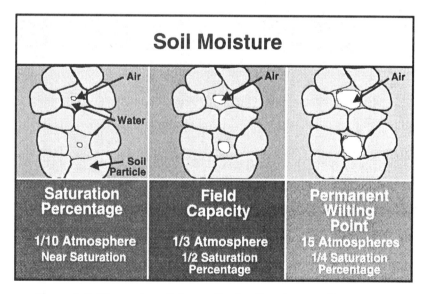

Figure 2-2. Illustration of the cardinal soil moisture values — saturation percentage, field capacity, and permanent wilting point.

four times the permanent wilting point. This relationship between the saturation percentage, field capacity, and the permanent wilting point is approximate for all soils from clay loams to sandy soils. Figure 2–2 schematically presents these relations.

A very moist soil has a thick film of water and hence has a low suction. A drier soil has a thin film of water and has a high suction. For this reason, water will move from a wet soil to a drier soil, but the rate of such movement is slow. Table 2–1 shows the available water-holding capacity of different soils.

### Table 2–1
### Approximate Amounts of Available Water
### Held by Different Soils

| Soil Texture | Inches of Water Held per Foot of Soil | Max. Rate of Irrigation — Inches per Hour, Bare Soil |
|---|---|---|
| Sand | 0.5 – 0.7 | 0.75 |
| Fine sand | 0.7 – 0.9 | 0.60 |
| Loamy sand | 0.7 – 1.1 | 0.50 |
| Loamy fine sand | 0.8 – 1.2 | 0.45 |
| Sandy loam | 0.8 – 1.4 | 0.40 |
| Loam | 1.0 – 1.8 | 0.35 |
| Silt loam | 1.2 – 1.8 | 0.30 |
| Clay loam | 1.3 – 2.1 | 0.25 |
| Silty clay | 1.4 – 2.5 | 0.20 |
| Clay | 1.4 – 2.4 | 0.15 |

Plants consume, on the average, from 0.1 to 0.3 in. of rainfall or irrigation per day.

## *WHEN TO IRRIGATE*

Plant appearance is often used as a guide in determining when to irrigate. Symptoms such as slow growth, "bluish" color of leaves, and temporary afternoon wilting are signs of moisture stress. Usually, plants should be irrigated before these signs are conspicuous.

A soil tube or an auger can show depth of wetting and depletion of moisture. Experienced individuals can determine fairly accurately the available moisture by feeling soil samples taken with a tube or an auger. Though this technique is quite subjective, many individuals rely exclusively on this method for determining when to irrigate. Table 2-2 presents a guide for determining the approximate soil water content using the "feel method."

Another practice which has been in use for many years is the *gravimetric* procedure. This method involves placing weighed samples of moist soil and drying them in an oven to determine the percentage of moisture loss.

## SOIL WATER MEASUREMENTS

Moisture-measuring instruments such as tensiometers and electrical resistance blocks indirectly measure soil moisture and are now used extensively.

Properly placed, they can be useful guides to determine the moisture levels in the soil and to schedule irrigations. Figure 2-3 illustrates two moisture measuring instruments.

A *tensiometer* consists of a porous ceramic cup imbedded in the major root zone of the soil. A rigid tube is connected to the cup and to a vacuum gauge just above the soil surface. The whole system is filled with water and sealed tightly. As the roots remove water from the soil, the soil draws water from the porous cup, creating a suction which is measured by the gauge. The drier the soil, the greater the suction. Tensiometers are effective for suctions from 0 to 0.8 atmospheres (80 centibars on most gauges). They are especially well suited for shallow rooted, frequently irrigated crops. Their primary limitation is ineffectiveness on dry soils (above 80 centibars).

Electrical resistance blocks consist of one or more pairs of

Table
Soil, Moisture, Appearance,

| Available Water[1] | Feel or Appearance | |
|---|---|---|
| | Sand | Sandy Loam |
| Above field capacity | Free water appears when soil is bounced in hand. | Free water is released with kneading. |
| 100% (field capacity) | Upon squeezing, no free water appears on soil, but wet outline of ball is left on hand. (1.0) | Appears very dark. Upon squeezing, no free water appears on soil, but wet outline of ball is left on hand. Makes short ribbon. (1.5) |
| 75–100% | Tends to stick together slightly, sometimes forms a weak ball with pressure. (0.8 to 1.0) | Quite dark. Forms weak ball, breaks easily. Will not slick. (1.2 to 1.5) |
| 50–75% | Appears to be dry, will not form a ball with pressure. (0.5 to 0.8) | Fairly dark. Tends to ball with pressure but seldom holds together. (0.8 to 1.2) |
| 25–50% | Appears to be dry, will not form a ball with pressure. (0.2 to 0.5) | Light colored. Appears to be dry, will not form a ball. (0.4 to 0.8) |
| 0–25% (0% is permanent wilting) | Dry, loose, single-grained, flows through fingers. (0 to 0.2) | Very slight color. Dry, loose, flows through fingers. (0 to 0.4) |

[1]Available water is the difference between field capacity and permanent wilting point.

[2]Numbers in parentheses are available water contents expressed as inches of water per foot of soil depth.

electrodes imbedded in a porous material, usually gypsum or fiberglass. The blocks are buried in the root zone with wires leading to the soil surface. As the soil water content changes, the moisture in the block changes with it. When the soil is wet, the electrical resistance is low. As the soil and block become dry, the resistance increases.

**2–2**
**and Description Chart**

**of Soil[2]**

| Loam/Silt Loam | Clay Loam/Clay |
|---|---|
| Free water can be squeezed out. | Puddles; free water forms on surface. |
| Appears very dark. Upon squeezing, no free water appears on soil, but wet outline of ball is left on hand. Will ribbon about 1 inch. (2.0) | Appears very dark. Upon squeezing, no free water appears on soil, but wet outline of ball is left on hand. Will ribbon about 2 inches. (2.5) |
| Dark color. Forms a ball, is very pliable, slicks readily if high in clay. (1.5 to 2.0) | Dark color. Easily ribbons out between fingers, has slick feeling. (1.9 to 2.5) |
| Fairly dark. Forms a ball, somewhat plastic, will sometimes slick slightly with pressure. (1.0 to 1.5) | Fairly dark. Forms a ball, ribbons out between thumb and forefinger. (1.2 to 1.9) |
| Light colored. Somewhat crumbly, but holds together with pressure. (0.5 to 1.0) | Slightly dark. Somewhat pliable, will ball under pressure. (0.6 to 1.2) |
| Slight color. Powdery, dry, sometimes slightly crusted, but easily broken down into powdery condition. (0 to 0.5) | Slight color. Hard, baked, cracked, sometimes has loose crumbs on surface. (0 to 0.6) |

This resistance is read by a specially designed meter attached to the wires. The meter is small and easily portable and can be used to monitor many stations. Resistance blocks operate effectively in the drier range of soil water content. Resistance blocks are limited by inaccuracy in highly saline soils, and by the need to replace blocks on a periodic basis (2-3 years).

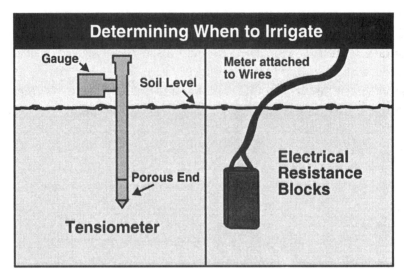

Figure 2-3. Two moisture-measuring and -monitoring instruments.

Neutron probes contain a radioactive source, a detector tube, and an electronic indicator unit. The source and detector tube form a single unit that is lowered into the soil through an access tube. Neutrons emitted by the source slow down as they collide with hydrogen atoms in the soil. The wetter the soil, the higher the count of slow neutrons. The returning slow neutrons are counted by the detector unit which then presents the data on a display or stores it in a minicomputer for future use. Neutron probes give accurate soil water content data over the entire root zone. When used with field capacity and permanent wilting point values, the data generated by the neutron probe allow the irrigation scheduler to determine when the soil has been depleted to a given percentage of available soil water. Potential disadvantages to the neutron probe include high cost, requires a trained and licensed operator, and diligent calibration to the sites being measured.

Thermal dissipation sensors are new soil-based instruments.

This device indirectly measures soil water content by measuring dissipation of heat from a porous ceramic block or disk in contact with the soil. The thermal dissipation sensor is well suited where high frequency irrigation is practiced. Among its many strengths are accuracy independent of soil texture, temperature, and salinity. Thus, once the sensor is calibrated, it is useful in any soil. It is also relatively maintenance free. Its weaknesses include high cost and the requirement for careful calibration of each sensor.

Time domain reflectometry is another new soil-based instrument. The TDR generates a high speed microwave pulse that travels down a transmission line buried in the soil. The velocity of the pulse in soils is determined primarily by water content. The returning pulse is detected, and via microprocessor, the travel time of the wave is used to calculate the water content of the soil.

Pressure bombs are devices used to measure plant moisture stress. The pressure bomb consists of a pressure chamber, pressure gauge, control valve, and a small tank of compressed nitrogen that serves as the pressure source. The petiole from a freshly cut leaf is extended through a rubber sealing stopper with the leaf inside of the pressure chamber. Upon cutting the petiole, plant fluids withdraw into the xylem vessel because the pressure outside the plant is several times greater than inside the conductive tissue. Once the chamber is sealed, pressure is slowly increased until water in the xylem is forced back out to the cut surface. At this point, the positive chamber pressure matches the negative potential of the xylem fluid. A higher gauge reading means a lower leaf water potential, which indicates increasing plant water stress. Disadvantages to the pressure bomb include measurement sensitivity to time of day, time required to take readings, and lack of leaf water potential guidelines for specific crops. Pressure bombs have gained greatest popularity in the cotton producing area of California's Central Valley.

Infrared thermometry indirectly measures plant moisture

stress by comparing crop canopy temperatures to ambient temperatures, relative humidity and sunlight intensity. Plants warm as they undergo moisture stress due to reductions in transpiration. Infrared thermometry measures canopy temperature and presents this information as stress indices. Changes in plant stress indices over time can be useful for evaluating the need to irrigate. Infrared thermometers are very portable, operate independent of soil conditions, and require very little preparation, calibration or maintenance. Leaf temperatures can be identified over a large area or for a single plant, depending on the thermometers distance from the crop and height above the canopy. The device is limited by its extreme sensitivity to user technique. Small differences in procedure can create very significant differences in plant measurements. In addition, each crop type requires development of specific stress indices for determining when to irrigate.

Although moisture sensing devices are useful, all have inherent limitations that must be thoroughly understood by the user. Future research is needed to refine the techniques and guidelines to make these tools of greater widespread value.

# WATER ANALYSIS
# TERMINOLOGY

Various terms are used in reporting chemical analysis of water. An understanding of the commonly used methods of reporting is needed to properly interpret the data.

Dissolved salts are dissociated into electrically charged particles called *ions*. The ions which are positively charged are called *cations*, while the negatively charged ions are called *anions*. The concentration of each of these types of ions in water is the usual method of reporting.

The common cations reported by laboratories are calcium ($Ca^{++}$), magnesium ($Mg^{++}$), sodium ($Na^+$), and potassium ($K^+$).

The anions usually reported are bicarbonate ($HCO_3^-$), carbonate ($CO_3^-$), chloride ($Cl^-$), and sulfate ($SO_4^-$).

An anion usually found in smaller amounts is nitrate ($NO_3^-$), which is sometimes reported as nitrate-nitrogen ($NO_3$-N).

Four principal units are used to express the concentration of constituents in water. They are grains per gallon, parts per million, milligrams per liter, and milliequivalents per liter.

Grains per gallon uses the English system of units. Most laboratories no longer use this method of reporting for irrigation water. It is still used, however, to report hardness of water. To change grains per gallon to parts per million, multiply grains per gallon by 17.2.

Part per million (ppm) is defined as 1 part of a salt to 1 million parts of water – or 1 milligram of salt per kilogram of solution. Since a kilogram of water equals 1 liter, ppm is interchangeable with *milligrams per liter (mg/l)*. Parts per million, or milligrams per liter, are often used to report boron, nitrate, nitrate-nitrogen, and a few other constituents measured in irrigation water.

Milliequivalents per liter (meq/l) is the most meaningful method of reporting the major chemical components of water.

Salts are combinations of cations (sodium, calcium, magnesium, etc.) and anions (chloride, sulfate, bicarbonate, etc.) in definite weight ratios. These weight ratios are based upon the atomic weight of each constituent and upon the valence (electrical charge). An equivalent weight of an ion is the atomic weight divided by the valence (see Table 2-3 for equivalent weights of the common ions). Thus, meq/l is a measurement of charge concentration/liter.

Due to the differences between various laboratories, it is useful to know how to convert from one unit of measurement to another. Parts per million, milligrams per liter and milligrams per kilogram can be used interchangeably and thus are identical for interpretive purposes. To convert ppm to meq/l (or vice versa) first determine the equivalent weight of the ion in question

Table 2–3
Major Constituents in Irrigation Water

|  | Ion Name | Symbol | Equivalent Weight |
|---|---|---|---|
|  |  |  | (g) |
| Cations | Calcium | Ca++ | 20 |
|  | Magnesium | Mg++ | 12 |
|  | Sodium | Na+ | 23 |
|  | Potassium | K+ | 39 |
| Anions | Bicarbonate | HCO$_3^-$ | 61 |
|  | Carbonate | CO$_3^=$ | 30 |
|  | Chloride | Cl$^-$ | 35.5 |
|  | Sulfate | SO$_4^=$ | 48 |

from Table 2–3. Then, follow these simple calculations for converting one unit of measurement to another:

$$\text{ppm} = \text{equivalent weight} \times \text{meq/l}$$
$$\text{meq/l} = \text{ppm} \div \text{equivalent weight}$$

Table 2–4 is useful in determining the actual amounts of dissolved salt applied to land per acre-foot of irrigation water. This calculation may be particularly useful for nitrate, and possibly others, in estimating the fertilizer or salt inputs derived from the water alone. For example, if water contains 7 ppm NO$_3$-N, then the amount of N applied per acre-foot = 7 × 2.72 = 19.0 lb N. If 3 acre-feet of water is applied during the season, then a total of 19.0 × 3 = 57 lb N is applied per season from the irrigation water used. From this sort of information, adjustments in the fertilizer program should be made.

The *pH* expresses the acidity or alkalinity of water. A pH reading of less than 7.0 is acidic, 7.0 is neutral, and above 7.0 is alkaline, or basic. Most well waters range from pH 7.0 to pH 8.5. Some stream waters may be as acidic as pH 5.5.

Total salt content is usually reported as the *electrical conductivity (EC$_W$)*. Chemically pure water does not conduct elec-

## Table 2–4
### Easy Conversion of Water Analysis Ions to
### Pounds Material per Acre-Foot of Water

| Material[1] | Pounds per Acre-Foot Multiplier[2] | |
|---|---|---|
| | *(ppm)* | *(me/l)* |
| $NO_3$ | 0.612 | 37.9 |
| $NO_3$-N | 2.72 | — |
| P | 6.23 | — |
| K | 3.26 | 127 |
| Ca | 2.72 | 54.4 |
| Mg | 2.72 | 32.9 |
| $CO_3$ | 2.72 | 81.6 |
| $HCO_3$ | 2.72 | 166 |
| Cl | 2.72 | 96.3 |
| $SO_4$ | 0.91 | 131 |
| Na | 2.72 | 61.7 |
| B | 2.72 | — |

[1]For P, K, and $SO_4$ the multiplier values listed incorporate conversion to the units of fertilizer expression $P_2O_5$, $K_2O$, and S respectively.

[2]Multiply the value presented in the water analysis report by the appropriate multiplier to determine the approximate pounds of material carried in an acre-foot of irrigation water.

tricity, but water with salts dissolved in it does. The more salt in the water, the better the conductor it becomes. The ability of a water sample to conduct electricity is used to determine the salt content. $EC_W$ is usually reported as decisiemens per meter (dS/m) or millimhos per centimeter (mmhos/cm). Both are equivalent units of measurement.

The total salt content or *total dissolved solids (TDS)* is usually reported as ppm. This can be determined by evaporating a known weight of water sample to dryness and weighing the salt remaining. More often it is estimated by measuring the $EC_W$ in dS/m (or mmhos/cm) and multiplying by 640. This estimates total dissolved solids in ppm. For example, if a water has an

$EC_W$ of 1.15, then 1.15 × 640 = 736 ppm total dissolved solids. To calculate the total pounds of dissolved salts applied per acre-foot of irrigation water, multiply ppm TDS by 2.72 (an acre-foot of water weighs 2.72 million pounds). Thus, water having an $EC_W$ of 1.15 would apply 736 × 2.72 = 2000 pounds of dissolved salt per acre foot of water! From this example it is easy to understand how salts in irrigation water contribute to the salinity of soils.

The TDS in me/l can be closely estimated by multiplying the $EC_W$ in dS/m by 10. For example, a water having an $EC_w$ of 2.62 dS/m contains 26.2 me/l of dissolved salts.

Electrical conductivity is commonly used to check the salt content of soils. The conductivity is measured on the soil saturation extract and is designated $EC_e$. This is used to monitor changes in salt concentration of the soil resulting from irrigation. It is also useful in evaluating the relative tolerance of plants to salt and the suitability of a soil for certain crops.

Percent sodium is sometimes used. It is the ratio of sodium to the total cations in milliequivalents. A high sodium percentage may indicate a poor quality water, but in recent years more refinement in interpretation of water quality has made this measurement obsolete.

# SOIL PROPERTIES AND WATER QUALITY

Over time, the quality of irrigation water and the irrigation practice used affects the characteristics of soil. These potential changes in soil properties may influence crop growth and performance.

## THE CATIONS

The major cations present in most irrigation waters, calcium, magnesium, and sodium, can have a profound impact on the physical as well as chemical properties of soils.

Calcium ($Ca^{++}$) is found in almost all natural waters. A soil well supplied with exchangeable calcium is friable, easily tilled and usually permits water to penetrate easily. Normally such soils do not "puddle" when wet. For these reasons, calcium in the form of gypsum is often applied to "tight" soils to improve their physical properties. Calcium replaces the sodium on the soil particles and allows the sodium to be leached below the root zone. Irrigation water containing predominantly calcium is most desirable.

Magnesium ($Mg^{++}$) is also found in measurable amounts in most natural waters. Magnesium behaves much like calcium in the soil. Often laboratories will not separate calcium and magnesium, but will report simply Ca + Mg in me/l. For most purposes, this is adequate.

Sodium ($Na^+$) salts are all very soluble and as a result generally are found in all natural waters.

A soil with a large amount of sodium associated with the clay fraction has poor physical properties for plant growth. When it is wet, it runs together, becomes sticky, and is nearly impervious to water. When it dries, hard clods form, making it difficult to till. Continued use of water with a high proportion of sodium may bring about these changes in an otherwise good soil. Eventually, a sodic soil may develop. The so-called "slick spots" are usually spots high in exchangeable sodium.

Potassium ($K^+$) is usually found in only small amounts in natural waters. It behaves much like sodium in the soil, but unlike sodium it is rarely associated with soil physical problems. It is most often reported separately in water analysis, but may be combined with sodium.

## THE ANIONS

The anions influence soil properties indirectly by affecting exchangeable calcium and sodium ratios, and directly by increasing salinity. The important anions are bicarbonate, carbonate, chloride, and sulfate.

Bicarbonate ($HCO_3^-$) is common in natural waters. It is not usually found in nature except in solution. Sodium and potassium bicarbonates can exist as solid salts. An example is baking soda (sodium bicarbonate). Calcium and magnesium bicarbonates exist only in solution. As the moisture in the soil is reduced, either by removal by plants or evaporation, calcium bicarbonate decomposes, carbon dioxide ($CO_2$) gas and water ($H_2O$) are formed, leaving insoluble lime ($CaCO_3$) behind.

Bicarbonate ions in the soil solution will precipitate calcium

$$Ca(HCO_3)_2 \xrightarrow{\text{upon drying}} CaCO_3 + CO_2 + H_2O$$

as the soil approaches dryness. This removes calcium from the clay and leaves sodium in its place. In this way, a calcium-dominant soil can become a sodium-dominant (sodic) soil by the use of a high-bicarbonate irrigation water.

Carbonate ($CO_3^=$) is found in some water having a pH greater than 8.0. Since calcium and magnesium carbonates are relatively insoluble, the cations associated with high-carbonate waters are likely to be sodium and possibly a small amount of potassium. When the soil dries, the carbonate ion will react with calcium and magnesium from the clay similar to bicarbonate, and an alkali (sodic) soil will develop.

Chloride ($Cl^-$) is found in all natural waters. In high concentrations, it is toxic to some plants. All common chlorides are soluble and contribute to the total salt content (salinity) of soils. The chloride content must be determined to properly evaluate irrigation water.

Sulfate ($SO_4^=$) is abundant in nature. Sodium, magnesium and potassium sulfates are readily soluble. Calcium sulfate (gypsum) has limited solubility. Sulfate has no characteristic action on the soil except to contribute to the total salt content. The presence of soluble calcium will limit its solubility.

Nitrate ($NO_3^-$) is rarely found in large amounts in natural water. Small amounts can affect the use of irrigation water by supplying plants with needed nitrogen, or in some cases with more than the desired amount of this plant nutrient. Large amounts of nitrate in water may indicate contamination from natural deposits and from other sources such as animal wastes, sewage, misuse of nitrogen fertilizers, and soil organic matter. Nitrate, however, has no significant effect on the physical properties of soil.

Boron (B) occurs in water in various anion and neutral forms. The usual range in natural water is from 0.01 to 1.0 ppm. Concentrations greater than this are known, but are most often from hot springs or brines.

Boron has no measurable effect on the physical properties of soil nor on soil salinity in the amounts that can be tolerated by plants. Boron is not as readily leached from the soil as chloride or nitrate, but most of it can be removed by successive leachings.

A small amount of boron is essential for plant growth, but concentration above the optimum may be phytotoxic. Some plants are more sensitive to an excess than others. Plants grown on sandy soils which have been irrigated by water exceedingly low in boron (less than 0.02 ppm) may develop boron deficiency.

## *EVALUATING IRRIGATION WATER*

A useful evaluation of irrigation water describes its effect on soils and plant growth. This effect is primarily related to the dissolved salts in the water. Depending upon the amount and kind of salts, different soil problems may develop. See Table 2-5 for guidelines to water quality interpretation.

A *saline soil* contains soluble salts in such quantities that they interfere with plant growth.

Table 2-5
Guidelines for Interpretation of Water Quality for Irrigation[1]

| Potential Problem | Units | Degree of Restriction on Use | | |
|---|---|---|---|---|
| | | None | Sl. to Mod. | Severe |
| pH | | Normal range 6.5 – 8.4 | | |
| Salinity | | | | |
| $EC_w$ (or) | dS/m | <0.7 | 0.7 – 3.0 | >3.0 |
| TDS | mg/l | <450 | 450 – 2,000 | >2,000 |
| Infiltration[2] | | | | |
| SAR = 0-3   and $EC_w$ = | | >0.7 | 0.7 – 0.2 | <0.2 |
| SAR = 3-6   and $EC_w$ = | | >1.2 | 1.2 – 0.3 | <0.3 |
| SAR = 6-12  and $EC_w$ = | | >1.9 | 1.9 – 0.5 | <0.5 |
| SAR = 12-20 and $EC_w$ = | | >2.9 | 2.9 – 1.3 | <1.3 |
| SAR = 20-40 and $EC_w$ = | | >5.0 | 5.0 – 2.9 | <2.9 |
| Specific Ion Effects | | | | |
| Sodium[3] | | | | |
|   Surface irrigation | SAR | <3 | 3 – 9 | >9 |
|   Sprinkler irrigation | me/l | <3 | >3 | |
| Chloride[3] | | | | |
|   Surface irrigation | me/l | <4 | 4 – 10 | >10 |
|   Sprinkler irrigation | me/l | <3 | >3 | |
| Boron | mg/l | <0.7 | 0.7 – 3.0 | >3.0 |
| Bicarbonate[4] | me/l | <1.5 | 1.5 – 8.5 | >8.5 |

[1]Adapted from University of California Committee of Consultants 1974.

[2]At a given SAR, infiltration rate increases as water salinity increases. Evaluate the potential infiltration problem by SAR as modified by $EC_w$. Adapted from Rhoades 1977, and Oster and Schroer 1979.

[3]For surface irrigation, most tree crops and woody plants are sensitive to sodium and chloride; use the values shown. Most annual crops are not sensitive; use salinity tolerance tables. For chloride tolerance of selected fruit crops, See Table 2-13. With overhead sprinkler irrigation and low humidity (<30 percent), sodium and chloride may be absorbed through the leaves of sensitive crops (see Table 2-14).

[4]Applies to overhead sprinkling only.

A *sodic (alkali) soil* contains enough sodium adsorbed on the clay particles to interfere with plant growth. If a sodic soil is relatively free of soluble salts, it is called a non-saline sodic soil. If, in addition to being sodic, it has sufficient soluble salts to restrict plant growth, it is called a saline-sodic soil.

Salinity hazard — One of the hazards of irrigated agriculture is the possible accumulation of soluble salts in the root zone. Some plants tolerate more salts than others, but all plants have a maximum tolerance. Most crops are more sensitive during early seedling growth and then become increasingly tolerant during later stages of growth and development. See Table 2-6 for examples of crop emergence as affected by soil salinity. With reasonably good irrigation practices, the salt content of the saturation extract of soil is about 1.5 to 3 times the salt content of the irrigation water. Where ample water is used to remove excess salt from the root zone, the salt level in the saturation extract is about 1.5 times that of the irrigation water. Where

Figure 2-4. Salts accumulate in the soil with inadequate leaching and drainage.

## Table 2–6
### Relative Salt Tolerance of Various Crops at Emergence[1]

| Crop | ECₑ for 50% Reduction in Emergence[2,3] |
|------|------|
| Barley | 16–24 |
| Cotton | 15 |
| Sugarbeet | 6–12 |
| Sorghum | 13 |
| Safflower | 12 |
| Wheat | 14–16 |
| Beet, red | 14 |
| Cowpea | 16 |
| Alfalfa | 8–13 |
| Tomato | 8 |
| Cabbage | 13 |
| Corn | 21–24 |
| Lettuce | 11 |
| Onion | 6–8 |
| Rice | 18 |
| Bean | 8 |

[1]Adapted from E.V. Maas, USDA Salinity Laboratory, Riverside, CA.

[2]$EC_e$ means electrical conductivity of the saturation extract of the soil reported in dS/m at 25°C.

[3]Emergence percentage of saline treatments determined when non-saline control treatments attained maximum emergence.

water is used more sparingly, there may be three times as much salt in the saturation extract.

With ordinary irrigation methods there is some leaching so the accumulation of salts in the soil is reduced but not eliminated. Before a critical assessment of the salinity hazard of any irrigation water is made, it is necessary to know how much salt the crop can tolerate and how much leaching is needed to reduce the salt in the soil to an acceptable level.

Tables 2-7 through 2-10 show the tolerance and leaching requirements estimated for numerous crops. Leaching need not be done at each irrigation, but should be done at least once a year.

Growers rotating crops must provide enough leaching so that damage to the most salt-sensitive crop in the rotation will be at a minimum.

With reasonable irrigation practices, there should be no salinity problems with irrigation water with an $EC_W$ of less than 0.75 dS/m. Increasing problems can be expected between $EC_W$ 0.75 and 3.0 dS/m. An $EC_W$ greater than 3.0 will cause severe problems, except for a few salt-tolerant crops.

It is generally assumed that the effects of saline water can be offset by increasing the amount of leaching so that the *average* salt content of the root zone is not increased. Plants respond to the average root zone salinity. For a given situation, the lower the leaching fraction, the higher the average root zone salinity.

Sodium or *permeability hazards* — In most cases, the permeability of the soil to water becomes a hazard before sodium has a toxic effect on plant growth. In a few plants, notably avocados, this may not be true.

As the proportion of sodium adsorbed on the clay increases, the soil tends to disperse or "run together" bringing about reduced rates of water penetration. The sodium adsorption ratio (SAR) indicates the relative activity of sodium ions as they react with clay. From the SAR the proportion of sodium on the clay can be estimated when an irrigation water has been used for a long period of time with reasonable irrigation practices.

Most laboratories will report the SAR of irrigation water. If not, it can be determined by using the following equation:

$$SAR = \frac{Na}{\sqrt{(Ca + Mg) \div 2}}$$

The sodium (Na), calcium (Ca), and magnesium (Mg) are expressed in me/l.

SAR is a good index of the sodium permeability hazard if

| Crop | $EC_e$[2] | $EC_w$[3] | LR | $EC_e$ | $EC_w$ | LR |
|------|------|------|------|------|------|------|
|      | *(0%)* | | | *(10%)* | | |
| Barley[5] | 8.0 | 5.3 | 9% | 10.0 | 6.7 | 12% |
| Cotton | 7.7 | 5.1 | 9% | 9.6 | 6.4 | 12% |
| Sugar beets[6] | 7.0 | 4.7 | 10% | 8.7 | 5.8 | 12% |
| Wheat[5, 7] | 6.0 | 4.0 | 10% | 7.4 | 4.9 | 12% |
| Safflower | 5.3 | 3.5 | 12% | 6.2 | 4.1 | 14% |
| Soybeans | 5.0 | 3.3 | 16% | 5.5 | 3.7 | 18% |
| Sorghum | 4.0 | 2.7 | 8% | 5.1 | 3.4 | 9% |
| Rice (paddy) | 3.0 | 2.0 | 9% | 3.8 | 2.6 | 11% |
| Corn | 1.7 | 1.1 | 6% | 2.5 | 1.7 | 8% |
| Flax | 1.7 | 1.1 | 6% | 2.5 | 1.7 | 8% |
| Cowpeas | 1.3 | 0.9 | 5% | 2.0 | 1.3 | 8% |
| Beans (field) | 1.0 | 0.7 | 5% | 1.5 | 1.0 | 8% |
| Broadbean | 1.6 | | | | | |
| Peanut | 3.2 | | | | | |
| Sugarcane | 1.7 | | | | | |

[1]Adapted from Quality of Water for Irrigation. R. S. Ayers. *Jour. of the Irrig. and Drain. Div.*, ASCE. Vol. 103, No. IR2, June 1977, p. 140, and Salt Tolerance of Plants. E. V. Maas. *Appl. Agri. Res.*, Vol. 1, No. 1, 1986, p. 12–26.

[2]$EC_e$ means electrical conductivity of the saturation extract of the soil reported in dS/m at 25°C.

[3]$EC_w$ means electrical conductivity of the irrigation water in dS/m at 25°C.

[4]Maximum $EC_e$ is the electical conductivity of the soil saturation extract at which crop growth ceases.

the water passes through the soil and reaches equilibrium with it. If the SAR is less than 6, there should be no problems with either sodium or permeability. In the range of 6 to 9, there are increasing problems; above 9, severe problems can be expected.

Reducing sodium related permeability problems may be accomplished by:

1. Applying a source of soluble calcium (gypsum) to the soil or directly in the irrigation water.

2–7
**Reduction in Yield[1]**

| $EC_e$ | $EC_w$ | LR | $EC_e$ | $EC_w$ | LR | $EC_e$[4] |
|--------|--------|-----|--------|--------|-----|-----------|
| (25%) | | | (50%) | | | (Maximum) |
| 13.0 | 8.7 | 16% | 18.0 | 12.0 | 21% | 28.0 |
| 13.0 | 8.4 | 16% | 17.0 | 12.0 | 22% | 27.0 |
| 11.0 | 7.5 | 16% | 15.0 | 10.0 | 21% | 24.0 |
| 9.5 | 6.4 | 16% | 13.0 | 8.7 | 22% | 20.0 |
| 7.6 | 5.0 | 17% | 9.9 | 6.6 | 23% | 14.5 |
| 6.2 | 4.2 | 21% | 7.5 | 5.0 | 25% | 10.0 |
| 7.2 | 4.8 | 13% | 11.0 | 7.2 | 20% | 18.0 |
| 5.1 | 3.4 | 15% | 7.2 | 4.8 | 21% | 11.5 |
| 3.8 | 2.5 | 12% | 5.9 | 3.9 | 20% | 10.0 |
| 3.8 | 2.5 | 12% | 5.9 | 3.9 | 20% | 10.0 |
| 3.1 | 2.1 | 12% | 4.9 | 3.2 | 19% | 8.5 |
| 2.3 | 1.5 | 12% | 3.6 | 2.4 | 18% | 6.5 |

[5]Barley and wheat are less tolerant during germination and seedling stage. $EC_e$ should not exceed 4 or 5 dS/m.

[6]Sensitive during germination. $EC_e$ should not exceed 3 dS/m for garden beets and sugar beets.

[7]Tolerance data may not apply to semi-dwarf varieties of wheat.

2. Reducing the pH and bicarbonate content of the irrigation water by adding sulfuric acid.

3. By incorporating sulfur in problem soils, provided adequate free lime is present. See Table 2–11.

The adjusted SAR ($SAR_a$) presented in previous editions of this handbook, and still reported by many laboratories, is no

**Table**
Vegetable Crops —

| Crop | $EC_e$ | $EC_w$ | LR | $EC_e$ | $EC_w$ | LR |
|---|---|---|---|---|---|---|
| | (0%) | | | (10%) | | |
| Beets[2] | 4.0 | 2.7 | 9% | 5.1 | 3.4 | 11% |
| Broccoli | 2.8 | 1.9 | 7% | 3.9 | 2.6 | 10% |
| Tomatoes | 2.5 | 1.7 | 7% | 3.5 | 2.3 | 9% |
| Cantaloupes | 2.2 | 1.5 | 5% | 3.6 | 2.4 | 8% |
| Cucumbers | 2.5 | 1.7 | 8% | 3.3 | 2.2 | 11% |
| Spinach | 2.0 | 1.3 | 4% | 3.3 | 2.2 | 7% |
| Cabbage | 1.8 | 1.2 | 5% | 2.8 | 1.9 | 8% |
| Potatoes | 1.7 | 1.1 | 6% | 2.5 | 1.7 | 9% |
| Sweet corn | 1.7 | 1.1 | 6% | 2.5 | 1.7 | 9% |
| Sweet potatoes | 1.5 | 1.0 | 5% | 2.4 | 1.6 | 8% |
| Peppers | 1.5 | 1.0 | 6% | 2.2 | 1.5 | 9% |
| Lettuce | 1.3 | 0.9 | 5% | 2.1 | 1.4 | 8% |
| Radishes | 1.2 | 0.8 | 4% | 2.0 | 1.3 | 7% |
| Onions | 1.2 | 0.8 | 5% | 1.8 | 1.2 | 8% |
| Carrots | 1.0 | 0.7 | 4% | 1.7 | 1.1 | 7% |
| Beans | 1.0 | 0.7 | 5% | 1.5 | 1.0 | 8% |
| Celery | 1.8 | | | | | |
| Squash (scallop) | 3.2 | | | | | |
| Squash (zucchini) | 4.7 | | | | | |
| Turnip | 0.9 | | | | | |

[1]Adapted from Quality of Water for Irrigation. R. S. Ayers. *Jour. of the Irrig. and Drain. Div.*, ASCE. Vol. 103, No. IR2, June 1977, p. 140, and Salt Tolerance of Plants. E. V. Maas. *Appl. Agri. Res.*, Vol. 1, No. 1, 1986, p. 12–26.

[2]Sensitive during germination. $EC_e$ should not exceed 3 dS/m for garden beets and sugar beets.

longer recommended. Careful evaluation by researchers reveal that this procedure overstates the effects of bicarbonate and carbonate ($HCO_3^-$ and $CO_3^-$) on the resulting sodium hazard. If the $SAR_a$ is to be used, the value should be further adjusted by a 0.5 multiplication factor to more accurately define the

2–8
Reduction in Yield[1]

| $EC_e$ | $EC_w$ | LR | $EC_e$ | $EC_w$ | LR | $EC_e$ |
|---|---|---|---|---|---|---|
| | (25%) | | | (50%) | | (Maximum) |
| 6.8 | 4.5 | 15% | 9.6 | 6.4 | 21% | 15.0 |
| 5.5 | 3.7 | 14% | 8.2 | 5.5 | 20% | 13.5 |
| 5.0 | 3.4 | 14% | 7.6 | 5.0 | 20% | 12.5 |
| 5.7 | 3.8 | 12% | 9.1 | 6.1 | 19% | 16.0 |
| 4.4 | 2.9 | 14% | 6.3 | 4.2 | 21% | 10.0 |
| 5.3 | 3.5 | 12% | 8.6 | 5.7 | 19% | 15.0 |
| 4.4 | 2.9 | 12% | 7.0 | 4.6 | 19% | 12.0 |
| 3.8 | 2.5 | 13% | 5.9 | 3.9 | 20% | 10.0 |
| 3.8 | 2.5 | 13% | 5.9 | 3.9 | 20% | 10.0 |
| 3.8 | 2.5 | 12% | 6.0 | 4.0 | 19% | 10.5 |
| 3.3 | 2.2 | 13% | 5.1 | 3.4 | 20% | 8.5 |
| 3.2 | 2.1 | 12% | 5.2 | 3.4 | 19% | 9.0 |
| 3.1 | 2.1 | 12% | 5.0 | 3.4 | 19% | 9.0 |
| 2.8 | 1.8 | 12% | 4.3 | 2.9 | 19% | 7.5 |
| 2.8 | 1.9 | 12% | 4.6 | 3.1 | 19% | 8.0 |
| 2.3 | 1.5 | 12% | 3.6 | 2.4 | 18% | 6.5 |

effects of bicarbonate on calcium precipitation. Newer and more accurate calculations for estimating the effects of bicarbonate have been developed. As a result, SAR is more correctly being reported as $SAR_{adj}$. The standards are synonymous for interpretive purposes.

Table
Fruit and Nut Crops —

| Crop | $EC_e$ | $EC_w$ | LR | $EC_e$ | $EC_w$ | LR |
|------|------|------|------|------|------|------|
| | *(0%)* | | | *(10%)* | | |
| Date palms | 4.0 | 2.7 | 4% | 6.8 | 4.5 | 7% |
| Figs<br>Olives<br>Pomegranates } | 2.7 | 1.8 | 6% | 3.8 | 2.6 | 9% |
| Grapefruit | 1.8 | 1.2 | 8% | 2.4 | 1.6 | 10% |
| Oranges | 1.7 | 1.1 | 7% | 2.3 | 1.6 | 10% |
| Lemons | 1.7 | 1.1 | 7% | 2.3 | 1.6 | 10% |
| Apples<br>Pears } | 1.7 | 1.0 | 6% | 2.3 | 1.6 | 10% |
| Walnuts | 1.7 | 1.1 | 7% | 2.3 | 1.6 | 10% |
| Peaches | 1.7 | 1.1 | 8% | 2.2 | 1.4 | 11% |
| Apricots | 1.6 | 1.1 | 9% | 2.0 | 1.3 | 11% |
| Grapes | 1.5 | 1.0 | 4% | 2.5 | 1.7 | 7% |
| Almonds | 1.5 | 1.0 | 7% | 2.0 | 1.4 | 10% |
| Plums | 1.5 | 1.0 | 7% | 2.1 | 1.4 | 10% |
| Blackberries | 1.5 | 1.0 | 8% | 2.0 | 1.3 | 11% |
| Boysenberries | 1.5 | 1.0 | 8% | 2.0 | 1.3 | 11% |
| Avocados | 1.3 | 0.9 | 8% | 1.8 | 1.2 | 10% |
| Raspberries | 1.0 | 0.7 | 6% | 1.4 | 1.0 | 9% |
| Strawberries | 1.0 | 0.7 | 9% | 1.3 | 0.9 | 11% |

[1]Adapted from Quality of Water for Irrigation. R. S. Ayers. *Jour. of the Irrig. and Drain. Div.*, ASCE. Vol 103, No. IR2, June 1977, p. 142.

Irrigation water with a very low salt content may also present a water infiltration problem. In this case the addition of some salt, preferably a calcium source, would be helpful. There is evidence that all irrigation waters should have a minimum calcium content of 20 ppm (1.0 me/l), and $EC_w$ of at least 0.5 dS/m to prevent dispersion of the soil.

## 2-9
## Reduction in Yield[1]

| $EC_e$ | $EC_w$ | LR | $EC_e$ | $EC_w$ | LR | $EC_e$ |
|---|---|---|---|---|---|---|
| | (25%) | | | (50%) | | (Maximum) |
| 10.9 | 7.3 | 11% | 17.9 | 12.0 | 19% | 32.0 |
| 5.5 | 3.7 | 13% | 8.4 | 5.6 | 20% | 14.0 |
| 3.4 | 2.2 | 14% | 4.9 | 3.3 | 21% | 8.0 |
| 3.2 | 2.2 | 14% | 4.8 | 3.2 | 20% | 8.0 |
| 3.3 | 2.2 | 14% | 4.8 | 3.2 | 20% | 8.0 |
| 3.3 | 2.2 | 14% | 4.8 | 3.2 | 20% | 8.0 |
| 3.3 | 2.2 | 14% | 4.8 | 3.2 | 20% | 8.0 |
| 2.9 | 1.9 | 15% | 4.1 | 2.7 | 21% | 6.5 |
| 2.6 | 1.8 | 15% | 3.7 | 2.5 | 21% | 6.0 |
| 4.1 | 2.7 | 11% | 6.7 | 4.5 | 19% | 12.0 |
| 2.8 | 1.9 | 14% | 4.1 | 2.7 | 19% | 7.0 |
| 2.9 | 1.9 | 14% | 4.3 | 2.8 | 20% | 7.0 |
| 2.6 | 1.8 | 15% | 3.8 | 2.5 | 21% | 6.0 |
| 2.6 | 1.8 | 15% | 3.8 | 2.5 | 21% | 6.0 |
| 2.5 | 1.7 | 14% | 3.7 | 2.4 | 20% | 6.0 |
| 2.1 | 1.4 | 13% | 3.2 | 2.1 | 19% | 5.5 |
| 1.8 | 1.2 | 15% | 2.5 | 1.7 | 21% | 4.0 |

It is now more fully understood that water infiltration is a factor of both water salinity ($EC_w$) and sodium (SAR) concentrations. Water infiltration rates generally tend to increase with increases in water salinity, and decrease with either decreasing salinity or increasing sodium, as SAR. Figure 2-5 is useful for estimating the combined effects of water salinity and sodium concentrations (SAR) on the resulting infiltration rate of soils.

Table
Forage Crops —

| Crop | $EC_e$ | $EC_w$ | LR | $EC_e$ | $EC_w$ | LR |
|------|------|------|------|------|------|------|
|      | (0%) | | | (10%) | | |
| Tall wheat grass | 7.5 | 5.0 | 8% | 9.9 | 6.6 | 10% |
| Wheat grass (fairway) | 7.5 | 5.0 | 11% | 9.0 | 6.0 | 14% |
| Bermudagrass | 6.9 | 4.6 | 10% | 8.5 | 5.7 | 13% |
| Barley (hay)[2] | 6.0 | 4.0 | 10% | 7.4 | 4.9 | 12% |
| Perennial ryegrass | 5.6 | 3.7 | 10% | 6.9 | 4.6 | 12% |
| Birdsfoot trefoil, narrow leaf | 5.0 | 3.3 | 11% | 6.0 | 4.0 | 13% |
| Harding grass | 4.6 | 3.1 | 9% | 5.9 | 3.9 | 11% |
| Tall fescue | 3.9 | 2.6 | 6% | 5.8 | 3.9 | 8% |
| Crested wheat grass | 3.5 | 2.3 | 4% | 6.0 | 4.0 | 7% |
| Vetch | 3.0 | 2.0 | 8% | 3.9 | 2.6 | 11% |
| Sudan grass | 2.8 | 1.9 | 4% | 5.1 | 3.4 | 7% |
| Big trefoil | 2.3 | 1.5 | 10% | 2.8 | 1.9 | 13% |
| Alfalfa | 2.0 | 1.3 | 4% | 3.4 | 2.2 | 7% |
| Clover, berseem | 1.5 | 1.0 | 3% | 3.2 | 2.1 | 6% |
| Orchardgrass | 1.5 | 1.0 | 3% | 3.1 | 2.1 | 6% |
| Meadow foxtail | 1.5 | 1.0 | 4% | 2.5 | 1.7 | 7% |
| Clover, alsike, ladino, red, strawberry | 1.5 | 1.0 | 5% | 2.3 | 1.6 | 8% |
| Lovegrass | 2.0 | | | | | |
| Wheatgrass | 3.5 | | | | | |
| Wildrye | 2.7 | | | | | |

[1]Adapted from Quality of Water for Irrigation. R. S. Ayers. *Jour. of the Irrig. and Drain. Div.*, ASCE. Vol. 103, No. IR2, June 1977, p. 140, and Salt Tolerance of Plants. E. V. Maas. *Appl. Agri. Res.*, Vol. 1, No. 1, 1986, p. 12–26.

[2]Barley and wheat are less tolerant during germination and seedling stage. $EC_e$ should not exceed 4 or 5 dS/m.

## 2–10
## Reduction in Yield[1]

| $EC_e$ | $EC_w$ | LR | $EC_e$ | $EC_w$ | LR | $EC_e$ |
|---|---|---|---|---|---|---|
| | *(25%)* | | | *(50%)* | | *(Maximum)* |
| 13.3 | 9.0 | 14% | 19.4 | 13.0 | 21% | 31.5 |
| 11.0 | 7.4 | 17% | 15.0 | 9.8 | 22% | 22.0 |
| 10.8 | 7.2 | 16% | 14.7 | 9.8 | 22% | 22.5 |
| 9.5 | 6.3 | 16% | 13.0 | 8.7 | 22% | 20.0 |
| 8.9 | 5.9 | 16% | 12.2 | 8.1 | 21% | 19.0 |
| | | | | | | |
| 7.5 | 5.0 | 17% | 10.0 | 6.7 | 22% | 15.0 |
| 7.9 | 5.3 | 15% | 11.1 | 7.4 | 21% | 18.0 |
| 8.6 | 5.7 | 12% | 13.3 | 8.9 | 19% | 23.0 |
| 9.8 | 6.5 | 11% | 16.0 | 11.0 | 19% | 28.5 |
| 5.3 | 3.5 | 15% | 7.6 | 5.0 | 21% | 12.0 |
| 8.6 | 5.7 | 11% | 14.4 | 9.6 | 18% | 26.0 |
| 3.6 | 2.4 | 16% | 4.9 | 3.3 | 22% | 7.5 |
| 5.4 | 3.6 | 12% | 8.8 | 5.9 | 19% | 15.5 |
| 5.9 | 3.9 | 10% | 10.3 | 6.8 | 18% | 19.0 |
| 5.5 | 3.7 | 11% | 9.6 | 6.4 | 18% | 17.5 |
| 4.1 | 2.7 | 11% | 6.7 | 4.5 | 19% | 12.0 |
| | | | | | | |
| 3.6 | 2.4 | 12% | 5.7 | 3.8 | 19% | 10.0 |

Table

Amounts of Sulfur Containing Amendments Required for
Sodium, or to Increase the Calcium Content in

| Chemical<br>Name | Trade Name/<br>Composition | Tons Equal<br>to One Ton<br>of Sulfur |
|---|---|---|
| Sulfur | 100% S | 1.00 |
| Gypsum | 100% $CaSO_4 \cdot 2H_2O$ | 5.37 |
| Potassium<br>thiosulfate | KTS<br>25% $K_2O$, 17% S | 5.88[3, 4]<br>11.8[4] |
| Ammonium<br>thiosulfate | Thio-sul<br>12% N, 26% S | 2.52[4, 5]<br>7.70[4] |
| Ammonium<br>polysulfide | Nitro-sul<br>20% N, 40% S | 1.59[5, 6]<br>3.13[6] |
| Monocarbamide dihydrogen<br>sulfate/sulfuric acid | N-pHuric 10/55<br>US-10<br>10% N, 18% S | 3.40[4, 7]<br>5.56[4] |
| Sulfuric acid | 100% $H_2SO_4$ | 3.06[4] |

[1]Not all the calcium, or potassium, will replace exchangeable sodium. The exchange reactions do not go entirely to completion. The extent of completion depends on replacement energies of calcium, potassium, ammonium, and sodium, and the exchangeable sodium percentage (ESP). More amendment is required than shown in this column. The correction for incomplete reactions involving calcium from gypsum or lime ranges from 1.14 times the amount shown for an ESP (or SAR) of 15 after the reactions are completed to 1.32 for an ESP (or SAR) of 5 after reactions are completed (Oster and Frenkel, 1980).

[2]The amount of exchangeable sodium to replace is the product of the cation exchange capacity times the difference between the initial ESP/100 (or SAR/100) and the desired final ESP/100 (or SAR/100).

## 2–11
## Calcareous Soils to Replace 1 meq/100 g of Exchangeable Irrigation Water or Soil Solution by 1 meq/L

| Pounds[1] Required per Acre to Replace 1 meq/100 g of Sodium[2] in 6 Inches of Soil | Pounds[1] Required per Acre-foot of Water to Obtain 1 meq/L of Calcium | Chemical Reactions That Occur in Calcareous Soils |
|---|---|---|
| 321 | 43.6 | $2S + 3O_2 + 2CaCO_3 + 4NaX = 2CaX_2 + 2Na_2SO_4 + 2CO_2$ |
| 1,720 | 234 | $CaSO_4 + 2NaX = CaX_2 + Na_2SO_4$ |
| 1,890 | 256 | $K_2S_2O_3 + 2O_2 + CaCO_3 + 4NaX$ |
| 3,770 | 513 | $= CaX_2 + 2KX + 2Na_2SO_4 + CO_2$ |
| 807 | 110 | $(NH_4)_2S_2O_3 + 2O_2 + CaCO_3 +$ |
| 2,470 | 336 | $2NaX = CaX_2 + (NH_4)_2SO_4 + Na_2SO_4 + CO_2$ |
| 510 | 69.1 | $(NH_4)_2S_5 + 8O_2 + 4CaCO_3 +$ |
| 1,000 | 136 | $8NaX = 4CaX_2 + (NH_4)_2SO_4 + 4Na_2SO_4 + 4CO_2$ |
| 1,090 | 148 | $H_2NCONH_2 \bullet H_2SO_4 + CaCO_3 +$ |
| 1,780 | 242 | $2NaX = CaX_2 + H_2NCONH_2 + Na_2SO_4 + CO_2 + H_2O$ |
| 981 | 133 | $H_2SO_4 + CaCO_3 + 2NaX = CaX_2 + Na_2SO_4 + CO_2 + H_2O$ |

[3]One mole of potassium assumed to replace beneficially one mole of exchangeable sodium.

[4]One mole of thiosulfate ($S_2O_3^=$) or sulfuric acid ($H_2SO_4$) assumed to result in replacement of two moles of exchangeable sodium by calcium.

[5]One mole of ammonium assumed to result in replacement of two moles of exchangeable sodium by calcium. Nitrification of ammonium to nitrate in calcareous soils releases one mole of calcium per mole of ammonium.

[6]One mole of polysulfide ($S_5^=$) assumed to result in replacement of eight moles of exchangeable sodium by calcium.

[7]One mole of nitrogen in monocarbamide dihydrogen sulfate assumed to result in replacement of one mole of exchangeable sodium by calcium.

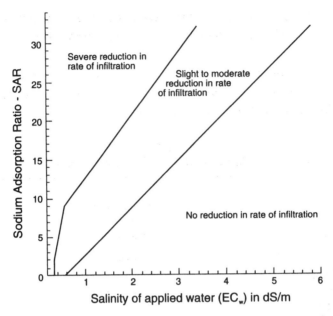

Figure 2-5. Relative rate of water infiltration as affected by salinity and sodium adsorption ratio (Adapted from Rhoades 1977; and Oster and Schroer 1979)

# TOXIC CONSTITUENTS

Several inorganic constituents such as boron, chloride, and sodium, may be found in natural water at levels which can be toxic to plants. High levels of bicarbonate in water have been shown to induce iron deficiency in some plants. This is minor when compared to the role of bicarbonates in creating permeability problems.

## BORON HAZARD

A small amount of boron is necessary for plant growth. To provide an adequate supply of this plant nutrient, 0.02 ppm B or more in the irrigation water may be required. Some water contains an adequate supply of boron, but water from many rivers may be deficient. Some well water and a few surface streams contain an excess of boron, thus creating a hazard by their use.

Plants grown in soils high in lime may tolerate more boron that those grown in non-calcareous soils. Table 2-12 presents the relative tolerance of plants to boron.

### Table 2–12
### Relative Tolerance of Plants to Boron in Water[1]

| Very sensitive (<0.5 ppm) | Artichoke, Jerusalem |
|---|---|
| Lemon[2] | Bean, kidney |
| Blackberry | Bean, lima |
|  | Peanut |
| **Sensitive (0.5–0.75 ppm)** | **Moderately sensitive (1.0–2.0 ppm)** |
| Avocado | Pepper, red |
| Grapefruit | Pea |
| Orange | Carrot |
| Apricot | Radish |
| Peach | Potato |
| Cherry | Cucumber |
| Plum |  |
| Persimmon | **Moderately tolerant (2.0–4.0 ppm)** |
| Fig, Kadota |  |
| Grape | Lettuce |
| Walnut | Cabbage |
| Pecan | Celery |
| Cowpea | Turnip |
| Onion | Bluegrass, Kentucky |
|  | Oats |
| **Less sensitive (0.75–1.0 ppm)** | Corn |
|  | Artichoke |
| Garlic | Tobacco |
| Sweet potato | Mustard |
| Wheat | Clover, sweet |
| Barley | Squash |
| Sunflower | Muskmelon |
| Bean, mung |  |
| Sesame | **Tolerant (4.0–6.0 ppm)** |
| Lupine |  |
| Strawberry | Sorghum |
|  | Tomato |

*(Continued)*

## Table 2–12 (Continued)

| | |
|---|---|
| Alfalfa | *Very tolerant* |
| Vetch, purple | *(>6.0 ppm)* |
| Parsley | Cotton (6.0–10 ppm) |
| Beet, red | Asparagus (10–15 ppm) |
| Sugar beet | |

[1]Adapted from Salt Tolerance of Plants E. V. Maas, in *Handbook of Plant Science.* 1984. Maximum permissible concentration in soil water without yield or vegetative growth reduction. Maximum concentrations in the irrigation water are approximately equal to these values or slightly less.

[2]Plants first named are considered more sensitive and last named are more tolerant.

## *CHLORIDE HAZARD*

Chloride ions are found in virtually all natural water. The chloride anion is soluble and moves freely through soil. Relatively small amounts of chloride are necessary for plant growth. However, high concentrations of chloride are toxic to some plants, particularly woody species.

Most annual crops and short lived perennials are moderately tolerant to chloride. Trees, vines, and woody ornamentals tend to be more sensitive. Table 2-13 lists the chloride tolerance for selected fruit crops and rootstocks.

### Table 2–13
### Chloride Tolerance of Some Fruit Crop Cultivars and Rootstocks[1]

| Crop | Rootstock or Cultivar | Maximum Permissible Cl⁻ without Leaf Injury[2] | |
|---|---|---|---|
| | | Root Zone ($Cl_e$) | Irrigation Water ($Cl_w$)[3,4] |
| | | *(me/l)* | *(me/l)* |
| | Rootstocks | | |
| Avocado | West Indian | 7.5 | 5.0 |
| *(Persea americana)* | Guatemalan | 6.0 | 4.0 |
| | Mexican | 5.0 | 3.3 |

*(Continued)*

## Table 2–13 (Continued)

| Crop | Rootstock or Cultivar | Maximum Permissible Cl⁻ without Leaf Injury[2] | |
| --- | --- | --- | --- |
| | | Root Zone $(Cl_e)$ | Irrigation Water $(Cl_w)$[3, 4] |
| | | (me/l) | (me/l) |
| Citrus (*Citrus* spp.) | Sunki mandarin Grapefruit Cleopatra mandarin Rangpur lime | 25.0 | 16.6 |
| | Sampson tangelo Rough lemon Sour orange Ponkan mandarin | 15.0 | 10.0 |
| | Citrumelo 4475 Trifoliate orange Cuban shaddock Calamondin Sweet orange Savage citrange Rusk citrange Troyer citrange | 10.0 | 6.7 |
| Grape (*Vitis* spp.) | Salt Creek, 1613–3 | 40.0 | 27.0 |
| | Dog Ridge | 30.0 | 20.0 |
| Stone Fruits (*Prunus* spp.) | Marianna | 25.0 | 17.0 |
| | Lovell, Shalil | 10.0 | 6.7 |
| | Yunnan | 7.5 | 5.0 |
| | **Cultivars** | | |
| Berries (*Rubus* spp.) | Boysenberry | 10.0 | 6.7 |
| | Olallie blackberry | 10.0 | 6.7 |
| | Indian summer raspberry | 5.0 | 3.3 |

*(Continued)*

## Table 2-13 (Continued)

| Crop | Rootstock or Cultivar | Maximum Permissible Cl⁻ without Leaf Injury[2] | |
| | | Root Zone ($Cl_e$) | Irrigation Water ($Cl_w$)[3, 4] |
|---|---|---|---|
| | | *(me/l)* | *(me/l)* |
| **Grape** | Thompson seedless | 20.0 | 13.3 |
| (*Vitis* spp.) | Perlette | 20.0 | 13.3 |
| | Cardinal | 10.0 | 6.7 |
| | Black Rose | 10.0 | 6.7 |
| **Strawberry** | Lassen | 7.5 | 5.0 |
| (*Fragaria* spp.) | Shasta | 5.0 | 3.3 |

[1]Adapted from Salt Tolerance of Plants. E. V. Maas, in *Handbook of Plant Science.* 1984.

[2]For some crops, the concentration given may exceed the overall salinity tolerance of that crop and cause some reduction in yield in addition to that caused by chloride ion toxicities.

[3]Values given are for the maximum concentration in the irrigation water. The values were derived from saturation extract data ($EC_e$) assuming a 15–20 percent leaching fraction and $EC_e = 1.5\ EC_w$.

[4]The maximum permissible values apply only to surface irrigated crops. Sprinkler irrigation may cause excessive leaf burn at values far below these (see Tables 2-5 and 2-14).

## *SODIUM TOXICITY*

Trees, vines, and woody ornamentals may be sensitive to excessive sodium. As in the case of chlorides, annual crops are usually not affected directly by high sodium, except for the contribution to soil salinity. For water with SAR below 3, there should be no restrictions on use or plant growth. Problems increase as SAR values increase from 3 to 9, and above 9 they may be severe. In addition, severe soil permeability problems would also be expected where irrigation water has SAR values above 9.

# WATER FOR
# SPRINKLER IRRIGATION

Most water suitable for surface irrigation may be safely used for overhead sprinkler irrigation. There are, however, some exceptions.

Leaf burn caused by sodium and chloride absorption may occur when the rate of evaporation is high. Conditions such as low humidity, high temperature, and winds can increase the concentration of these ions in the water on the leaves. Sometimes this can be corrected by increasing the frequency of irrigation. If this is impractical, it may be necessary to irrigate only at night during periods of hot, dry weather. Usually there is no problem when the irrigation water contains 3 me/l or less of either sodium or chloride.

Certain crops, such as almond and other deciduous fruit and nut trees, are sensitive to foliar applied salts because they absorb sodium and chloride very readily through the foliage. Other crops which are very sensitive to soil salinity, such as avocado and strawberry, absorb salts very slowly through the foliage and are therefore very tolerant of salts applied to the foliage. Table 2-14 presents approximate sodium or chloride

### Table 2–14
### Tolerance of Selected Crops to Foliar Injury from Saline Water
### Applied by Overhead Sprinklers[1]

| Na or Cl Concentrations Causing Injury (me/l) | | | |
|---|---|---|---|
| < 5 | 5 – 10 | 10 – 20 | > 20 |
| Almond | Grape | Alfalfa | Cauliflower |
| Apricot | Pepper | Barley | Cotton |
| Citrus | Potato | Corn | Sugarbeet |
| Plum | Tomato | Cucumber | Sunflower |
|  |  | Safflower |  |
|  |  | Sesame |  |
|  |  | Sorghum |  |

[1]Adapted from Salt Tolerance of Plants. E. V. Maas. *Applied Agri. Res.*, Vol. 1, No. 1, 1986, p. 12–26.

concentrations that can induce toxicity when applied through overhead sprinklers.

Bicarbonate ions in water can also be a problem with overhead sprinkler irrigation. A white deposit of calcium carbonate may form on the leaves and fruit. This can render some fruits and ornamentals unmarketable because they are unattractive. This coat of "whitewash" is not known to have an adverse effect on plant growth. Bicarbonate levels below 1.5 me/l should cause no problem.

# LOW VOLUME IRRIGATION

Low volume irrigation is the frequent slow application of water through various types of emitters. The most common forms of low volume irrigation are drip and microsprinkler. Low volume irrigation system design, components, and flow rate requirements vary according to the crop, soil, water, and environmental conditions. For example, a mature tree may require from two to six emitters, with each emitter applying 1/2 to 2 gallons per hour.

These low rates of application require small orifices and therefore need water which has been filtered free of solid particles. Dissolved salts in the water may crystalize around the orifices and reduce their flow rates or plug them completely. The most common cause of chemical plugging in low volume irrigation systems is the formation of insoluble calcium carbonate. Other causes of plugging include chemical or microbial oxidation of iron or manganese, bacterial or algal growth, suspended solids, or reaction of injected fertilizers with ions present in the irrigation water. Careful attention to water quality tests is required in order to evaluate the plugging potential of irrigation water. Table 2-15 presents the plugging potential of water used for low volume irrigation.

The total amount of water used in low volume irrigation is usually less than in conventional irrigation systems because:

## Table 2-15
## Plugging Potential of Irrigation Water
## Used in Drip Irrigation Systems[1]

| Type of Problem | Potential Restrictions on Use | | |
|---|---|---|---|
| | Little | Slight to Moderate | Severe |
| **Physical** | | | |
| Suspended solids (mg/l) | <50 | 50 - 100 | >100 |
| **Chemical** | | | |
| pH | <7.0 | 7.0 - 8.0 | >8.0 |
| Dissolved solids (mg/l) | <500 | 500 - 2,000 | >2,000 |
| Manganese (mg/l) | <0.1 | 0.1 - 1.5 | >1.5 |
| Iron (mg/l) | <0.1 | 0.1 - 1.5 | >1.5 |
| Hydrogen sulfide (mg/l) | <0.5 | 0.5 - 2.0 | >2.0 |
| **Biological** | | | |
| Bacterial populations (maximum number/ml) | <10,000 | 10,000 - 50,000 | >50,000 |

[1]Adapted from Water Analysis and Treatment Techniques to Control Emitter Plugging. F. S. Nakayama. From *Proceedings Irrig. Assoc. Conf.* p. 21-24, Portland, OR. Feb. 1982.

1. Tailwater runoff can be completely eliminated.

2. Evaporation may be reduced because less of the soil surface is wetted.

3. Total volume of soil wetted is usually less.

4. Deep percolation of water may be reduced.

Low volume irrigation is more widely used on permanent crops such as orchards and vineyards, especially on irregular slopes. Since it may save water, it is also adapted to areas where water is costly or scarce.

Low volume irrigation systems are gaining in popularity on some annual crops where costs can be justified through enhanced yields or quality, or reduced water costs.

# CROP SALINITY TOLERANCE
# AND LEACHING REQUIREMENTS

The crop salinity tolerance tables (Tables 2-7 through 2-10) show:

1. Typical yield reduction due to salinity of irrigation water ($EC_w$).

2. Typical salt content of the soil saturation extract ($EC_e$), at a given $EC_w$, when common surface irrigation methods are used.

3. Leaching requirement (LR), which is the fraction of the irrigation water that must be leached through the active root zone to minimize soil salinity at a specific $EC_w$.

4. Maximum concentration of salts at which plant growth ceases ($EC_eMAX$).

These tables were developed using data from the USDA Salinity Laboratory, Riverside, California.

An example of how to use these tables:

1. Assume the crop is corn.

2. From Table 2-7:
   Maximum $EC_w$
       for 0% yield loss = 1.1 dS/m with LR 6%
       for 10% yield loss = 1.7 dS/m with LR 8%
       for 25% yield loss = 2.5 dS/m with LR 12%
       for 50% yield loss = 3.9 dS/m with LR 20%

Suppose the water has an $EC_W$ of 2.7 dS/m. Using common surface irrigation methods, one can expect the $EC_e$ of the most active root zone to be 1.5 times the $EC_W$ (2.7) or 4.0 dS/m. The table shows that a yield decrease of about 25 percent can be expected.

Some water must pass below the root zone to maintain the

$EC_e$ within safe levels. This minimum leaching requirement may be calculated by the following equation:

$$LR = \frac{EC_w}{5\,(EC_e) - EC_w}$$

where   $EC_w$ = measured $EC_w$ of the irrigation water, and
       $EC_e$ = estimated soil $EC_e$ corresponding to an acceptable yield potential (from Tables 2-7 to 2-10)

As an example, assume the crop is tomatoes, and the $EC_w$ of the irrigation water is 1.2 dS/m. The grower desires to achieve 0% yield reduction from salts. The soil salinity threshold for tomatoes is 2.5 dS/m (from Table 2-8).

$$\frac{1.2}{5\,(2.5) - 1.2} = \frac{1.2}{11.3} = 0.11$$

Thus, a minimum leaching requirement of 11%, in relation to the crops evapotranspiration requirement, must be applied to maintain soil salinity at safe levels.

Another, and perhaps simpler, method for estimating the required leaching fraction is presented in Figure 2-6. With this method, the desired $EC_e$ (from Tables 2-7 through 2-10) is divided by the $EC_w$ of the irrigation water. This produces the "Multiplication Factor." Next, read horizontally from the "Multiplication Factor" to the curved line, then downward to the "Leaching Fraction" scale. This is the amount of extra water, in relation to the crops evapotranspiration requirement, which must be added to maintain the soil salinity at the desired level. For example, assume the crop to be grown is *tomatoes,* and the $EC_w$ of the irrigation water is 1.2 dS/m. Tomato has a salinity threshold of 2.5 dS/m. Divide 2.5 by 1.2 for a Multiplication Factor of 2.08. On the Multiplication Factor scale, read over to the curved line and down to the Leaching Fraction scale and read 0.11. This means a minimum 11% leaching fraction of irrigation

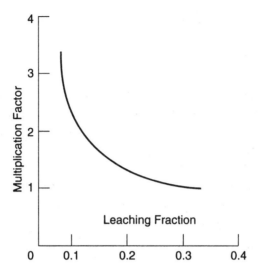

¹ Adapted from Oster and Rhoades, *Irrigation With Reclaimed Waste Water*,

- Divide Salinity Threshold of the crop in question by the $EC_w$ (Electrical Conductivity of Water) to arrive at the Multiplication Factor.

- On Multiplication Factor scale, read horizontally across to curved line then vertically to Leaching Fraction scale to read Leaching Fraction.

**Figure 2-6. Guidelines for determining leaching requirements.**

water should be applied to maintain a safe soil salinity level. Crop evapotranspiration (ET) data are available from the Extension Service. If tomatoes require 27 inches ET, then an additional 11%, or 3 inches, should be applied to meet the required leaching fraction. Thus, 30 inches should be applied during the season to satisfy both crop ET and leaching requirement.

A general rule of thumb for estimating the percent of soluble salts leached from the top one-foot of soil is as follows:

> 6" of water moving through the soil will leach approximately 50% of the salts;
> 12" of water moving through the soil will leach approximately 80% of the salts;
> 24" of water moving through the soil will leach approximately 90% of the salts.

# SALT MOVEMENT IN SOIL

Soluble salts in soil move with water. In areas of overall flooding or sprinkling, salt movement is directly downward. Upward and horizontal movement and concentration of salts on beds, border checks or berms commonly occurs.

Seedlings of most crops are more sensitive to soluble salts than are established plants. Planting on the shoulder of the bed or planting two rows on a wide bed so that the salts will be pushed to the center of the bed and away from the seeds or seedlings may help to prevent damage (see Figure 2-7). Sprinkling irrigated seedlings until they become well established may also be desirable.

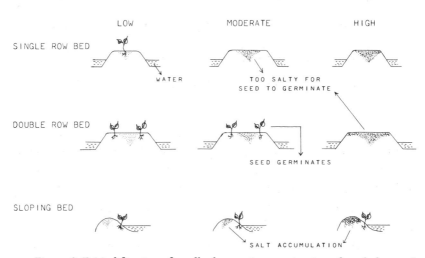

Figure 2-7. Modification of seedbeds permits germination of seeds for good stand establishment.

Excessive salts may accumulate in the tops of beds during pre-irrigation (see Figure 2-8). This is particularly true where animal manures have been used and subsequent rainfall has not been great enough to remove excess salt before planting the

PATTERNS OF SALT ACCUMULATION

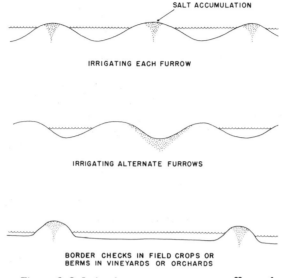

Figure 2-8. Irrigation management can affect salt accumulation.

seed. In this case, it may be necessary to sprinkle irrigate to remove the salt or to disperse the salts before seeding.

In areas of low rainfall, permanent berms in the rows of orchards and vineyards may accumulate excessive salts in a few years even when relatively low-salt water is used. In such cases, these berms should be removed to disperse the salt and then be rebuilt.

# *DRAINAGE*

Drainage may be necessary for soils with high water tables or where salinity is a hazard. Most crops grow best where the water table is more than 6 feet below the soil surface. Fields where water stands within 6 feet of the surface should be drained, if possible. This can be accomplished by the installation

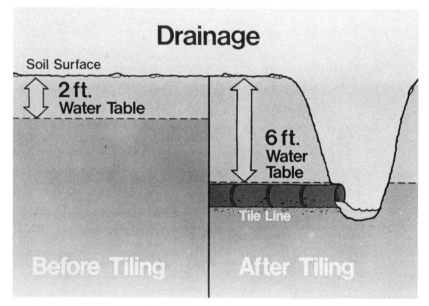

Figure 2-9. Tile drainage is used to lower the water table.

of a drainage system. Drains will remove water if placed below the water table or in the zone of saturation. If a field is poorly drained and there is an outlet for the drainage water, a carefully designed and installed drainage system may pay handsomely.

In certain areas, such as the Sacramento Delta and Lower Klamath Lake, crops may obtain a large share of their water needs from shallow ground water.

## SUPPLEMENTARY READING

1. Agricultural Salinity Assessment and Management. K. K. Tanji. ASCE Manuals and Reports on Engineering, Practice No. 71. ASCE Publ. NY. 1990.

2. Diagnosis and Improvement of Saline and Alkali Soils. USDA Agricultural Handbook No. 60, 1954.

3. Modern Irrigated Soils. D. W. James, R. J. Hanks and J. J. Jurinak. Wiley-Interscience. 1982.

4. Quality of Water for Irrigation. J. D. Rhoades. *Soil Sci.* 113:277-284. April 1972.

5. Salt Tolerance of Plants. E. V. Maas in *Handbook of Plant Science*. B. R. Christie, ed. CRC Press, Inc. 1984.

6. Tables for Calculating pHc Values of Waters. L. V. Wilcox. USDA Salinity Lab. Mimeo. December 1966.

7. Water Penetration Problems in California, California Soils: Prevention, Diagnosis and Solution. J. D. Oster, M. J. Singer, A. Fulton, W. Richardson, T. Prichard. Publ. Kearney Foundation of Soil Science, 1992.

8. Water Quality for Agriculture. R. S. Ayers and D. W. Westcott. FAO 29 rev. 1. NY. 1985.

9. Water Relations of Plants. P. J. Kramer, ed. Academic Press, Inc. 1983.

# Chapter 3

# *Principles of Plant Growth*

Webster defines growth as "a growing; increase; esp., progressive development of an organism, or the like." This is well illustrated by the process of planting a seed and initiating the process that results in the production of a whole plant. Plant growth is the process that provides people with their food supply and much of their shelter. Over a century ago the noted scientist Justus von Liebig said that plant growth is "the primary source whence man and animals derive the means of their growth and support."

Growth is one of the fundamental attributes of living organisms. It is more than a mere increase in size and weight, although this may be one expression of it. Growth represents a progressive and irreversible change in form involving the formation of new cells and their enlargement and maturation into the tissues and organs of the plant.

All plants must have, in varying degrees, the same basic supply of light, heat, energy, water, oxygen, carbon, and mineral elements for growth. Usually the soil supplies the needed moisture and mineral elements while the air supplies the oxygen and carbon dioxide. Growth is stopped, started, or at least modified as environmental conditions change, both in the root zone and around the aerial portion of the plants. This chapter will briefly discuss the requirements for satisfactory growth and introduce some of the concepts involved in the mineral nutrition of plants.

# THE PLANT CELL

All living plant parts are made up of cells. The single plant cell is the basic structural and functional unit of the plant. It is a tiny chemical factory that absorbs and secretes materials; transforms light energy into chemical energy (photosynthesis); respires and releases energy for various activities; digests or transforms foods; synthesizes complex chemicals from air, water, and simple sugars; and contains and even synthesizes the remarkable substance called protoplasm. These are but a few of the activities carried on within the cell which are basic to the life process itself. A generalized plant cell is shown in Figure 3-1.

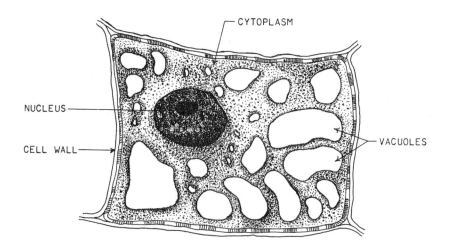

Figure 3-1. All living plant parts are made up of cells.

# PLANT TISSUES

Groups of cells that function as a unit are referred to as "tissues." Tissues may then be classified by function into one of four groups: meristematic, fundamental, protective, and vascular. Meristematic tissues are the embryonic and undifferentiated cells capable of growing by division which occur at growing points. Fundamental tissues are made up of masses of cells which have little specialization in structure or function. They act primarily as storage units. The epidermal or "skin" surface of a plant usually contains protective tissues. Vascular tissues such as xylem and phloem function in the conductive processes of the plant. These highly specialized tissues also add mechanical support because of their structure and location.

# PLANT ORGANS

The organization of a group of tissues forms an organ. Organs are generally separated into roots, leaves, stems, and reproductive structures. However, throughout the plant kingdom there are plants that may have one or more of these organs missing and other specialized organs present. A careful study of botany will reveal many interesting things about plant life and will contribute to one's understanding of its complexities.

# ROOTS

The root is that part of the plant which ordinarily grows downward into the soil, anchors the plant, and absorbs water and mineral nutrients. It may also serve as a food storage organ or reproductive organ, or it may perform some other function. Roots vary greatly in form and size among species. In a study of a single rye plant that was allowed to grow for four months in a box 12 inches square and 22 inches deep, it was found that the accumulated length of all the roots was 387 miles. The total area of the root surface of that one plant was 6,875 square feet.

The root differs from the shoot portion of the plant primarily in structure. Unlike stems, roots do not normally bear leaves or buds and are not divided into nodes and internodes. Usually, the root differs from the shoot in function and location, but this is not always so, since some plants have roots that develop buds which give rise to leafy shoots, and other plants have aerial stems which absorb water and nutrients. Tubers and stolons are stems often found underground, while brace roots of corn and air roots of orchids, for example, are found above ground.

Root systems are often grouped into two general types, fibrous and taproot. When numerous long, slender roots of about equal size are developed, they are known as fibrous roots. Examples of this root classification are corn, small grains, and the grasses. If the primary root remains the largest root of the plant and continues its downward development with other roots developing from it, it is classified as a taproot. Examples of this are cotton, alfalfa, sugar beet, and dandelion. Root systems not well defined as fibrous or taproot are also quite common. Whatever the classification, it is important to know the nature of the root system of a plant if one is to know how to properly manage the growing of that plant. How extensive is the root system? How deep are the roots? How rapidly does it develop? No plant can achieve its optimum development with a poor root system.

Numerous environmental factors influence root growth. These include light, gravity, temperature, salt concentration, soil texture and physical condition, oxygen supply, moisture, mineral nutrient supply, and overall plant health. Each root becomes subject to a combination of all of these factors, and its growth is the result of the combined action of all of them.

Roots develop best in friable and fertile soil. One of the purposes of tillage and fertilization is to provide a proper environment for the establishment and development of a good root system.

Oxygen must be available to all living cells. The amount necessary for growth varies with species. Unless the plant has specially developed systems for transporting oxygen to the roots (e.g., some of the swamp plants and rice), flooding of soils for appreciable lengths of time will cause root death because of lack of oxygen.

Downward growth is the common response of roots to the pull of gravity. Downward bending by the roots is known as positive geotropism, and upward bending by the shoots is negative geotropism.

The effect of light on hormonal concentration causes plants to bend toward the light. This is caused by unequal hormonal distribution due to light exposure. Bending toward light is termed positive phototropism; bending away from light is known as negative phototropism. Some roots exhibit negative phototropism.

Roots will generally grow in the direction of most favorable temperatures, thus exhibiting a positive thermotropism, and they will grow in the direction of favorable moisture supply (positive hydrotropism). These are merely responses to the immediate environment. Roots do not *seek* favorable temperatures or water, but grow in areas of favorable temperatures or moisture to which they are intimately exposed.

A close look at a root as illustrated in Figure 3-2 will show that it is made up of many different functioning parts. The primary absorbing region for water and mineral nutrients is the younger portion near the root tip. An even closer look in this region will reveal tiny root hairs radiating outward from all sides of the root. Root hairs (Figure 3-3) are about 0.01 millimeter in diameter and a few millimeters in length. Each hair is an outward extension of a portion of an epidermal cell. The outer, delicate walls of the root hair consist partly of pectic materials which are gelatinous and enable the root hair to cling to soil particles and to absorb water and salts in solution. It is common to find 200 to 300 root hairs per square millimeter of epidermis in the root-hair zone.

Proceeding from the growing tip back to the older root zone, the roots take on more the appearance and, to a degree,

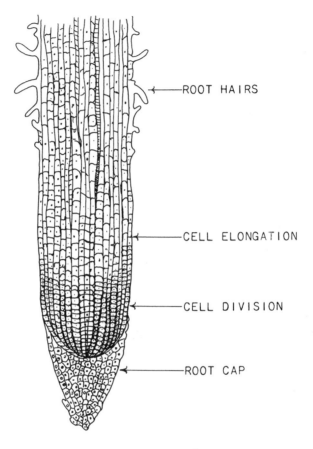

ROOT HAIRS

CELL ELONGATION

CELL DIVISION

ROOT CAP

Figure 3-2. A microscopic view of a root tip.

the functions of underground stems. Their primary role becomes one of conducting water and mineral nutrients from the absorbing root tip zone and transporting synthesized compounds from the leaves to the root tips.

A cross section of a typical root is shown in Figure 3-4. The outer layer of cells, called the epidermis, provides protection; the cortex gives support, and the stele contains the conductive tissues.

As a rule, most of the water and minerals the plant obtains are taken in by the younger roots in the newly developed root

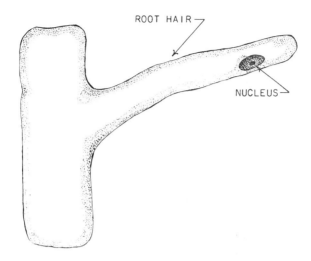

Figure 3-3. A close look at a root-hair cell.

area where the root hairs are most numerous. The older tissues back of this region become progressively impermeable although it has been shown that there may be considerable water absorption through these less active regions of the root, particularly in tree crops.

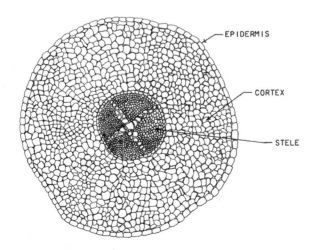

Figure 3-4. A cross section of a typical root.

The process of absorption is one of the main functions of the root. Without a constant supply of water, the plant cannot carry on the basic physiological activities such as photosynthesis, respiration, and growth. Without a supply of mineral nutrients brought in by absorptive processes, the plant would cease to live.

The study of plant physiology has brought out much concerning absorption of mineral nutrients. It has been shown to be a process requiring an expenditure of energy for certain nutrients. Energy is required to concentrate nutrient ions within the plant root. For example, ion concentrations may be 1,000 times greater within the cell than in the soil solution immediately outside the cell.

A brief, greatly simplified description of the absorption and movement of a mineral ion such as potassium will serve to illustrate the absorptive process. A typical young plant root cell consists of cytoplasm encircled on the outside by a cell wall (see Figure 3-1). Within the cytoplasm are vacuoles holding organic solutes and inorganic ions in solution. The membrane adjoining the cell wall is termed the *plasmalemma,* while another membrane termed the *tonoplast* forms the boundary between the vacuole and the cytoplasm. It is from the outer cell wall, through the plasmalemma, across the cytoplasm and the tonoplast, and into the vacuole, that the potassium ion must pass. Then for it to move into the conductive xylem tissue, it must traverse other cells and pass through additional membranes.

The potassium ion diffuses from the soil to the root surface. Here it freely moves through the cell wall as it is carried in the soil solution. At this point it contacts the plasmalemma. It is this membrane which is highly-impermeable to ions that controls the further movement of the potassium ion into the root. Certain locations or sites are found in this membrane which are specific for particular ions. At this point a carrier system attaches itself to the potassium ion, transports it across the plasmalemma and deposits it on the other side. The ion is then held inside by the membrane, and the carrier is regenerated to pick up another ion.

Once inside the cell, the potassium ion can then move by diffusion, by mass flow in the transpiration stream, or by processes regulated by metabolism. Entry into the upward-moving stream in xylem tissues allows for rather rapid movement of the potassium ions throughout the whole plant.

Water movement is primarily a physical process. As water evaporates from leaves it creates a difference in tension between the leaves and roots. This tension "pulls" water up the plant. Since the initial entry of water into the root is mainly through the active region of the root, into and through the cell walls, it resembles a long tube. The major impediment to the movement is through the endodermis where it must pass through cytoplasmic space. This is the area that is influenced by metabolism and which becomes subject to factors which affect metabolism such as temperature, oxygen supply, metabolic poisons, carbohydrate or food supply, etc.

Once water has entered the major conductive tissues, it travels with relatively minimal resistance to flow. These tissues are xylem and phloem. They are usually adjacent to one another and constitute the vascular tissues of the roots and the stems.

Numerous books have been written concerning such subjects as mineral nutrition of plants, plant and soil water relationships, and water relations of plants. Also, thousands of scientific articles have been written on these subjects. A few helpful references are included at the end of this chapter.

## *SHOOTS*

Leaves and stems make up shoots. Each of these structures has separate but often overlapping functions. A generalized cross section of a leaf is shown in Figure 3-5. The leaf is the center of photosynthetic activity — the initiation of the food manufacturing process. Its structure facilitates this function as it is physically positioned to get maximum exposure to the sun. Along its surface are minute openings called stomates, or stomata, where exchange of carbon dioxide and oxygen occurs. Underneath the surface or epidermal layer are specialized cells called the "spongy layer," which are ideally suited to perform the photosynthetic process. Here the gases from air, water in the cells, and light from the sun all join in the presence of chloroplasts, the chlorophyll-bearing bodies, to perform the photosynthetic process.

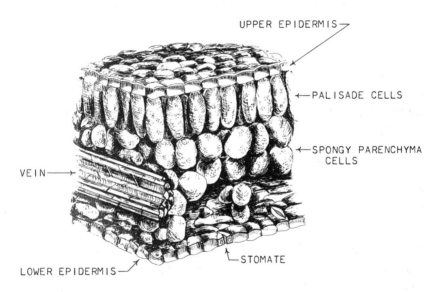

Figure 3-5. A cross section view of a leaf.

The stems of some plants also carry on photosynthesis, although these structures usually are the conductive tissues that carry water and mineral nutrients to the leaves and photosynthesized food to the roots or other organs. They also serve to support and display leaves and may become the food storage and reproductive units of the plant.

# PHOTOSYNTHESIS

The process of fixing carbon as six-carbon sugars from carbon dioxide and water is called photosynthesis. The energy required to form the sugar molecules from carbon, hydrogen, and oxygen is derived from light (sun or artificial). The reaction takes place within the chlorophyll molecule. The overall chemical equation for photosynthesis is:

$$6CO_2 + 12H_2O \xrightarrow[\text{chlorophyll}]{\text{light}} C_6H_{12}O_6 + 6O_2 + 6H_2O$$

carbon dioxide     water               sugar     oxygen    water

The process of respiration is just the reverse, or simply stated, the utilization of sugar in the presence of oxygen and water. In this case, light and chlorophyll are not needed.

The sugar produced by photosynthesis is transported to other plant parts. There it may be respired or stored after being converted to starch, fats, proteins, and other compounds. These stored compounds provide the animal world with its basic food supply.

Respiration, the consumptive oxidation of sugars, provides energy for many different metabolic reactions, the sum of which comprise the growth and "living" of the plant. Some of the reactions are organic compound production, such as amino acid synthesis, protein formation, and synthesis of fats and waxes. All of these essential reactions require the energy originally derived from light.

Most of the essential plant nutrients are utilized in or by these reactions (e.g. nitrogen in amino acids and proteins, magnesium and nitrogen in chlorophyll, phosphorus in the ADP-ATP reaction). Specific functions of the nutrients are covered in Chapter 4 of this book.

## *TRANSPIRATION*

The evaporation of water from leaves, stems, and other aerial parts of plants is called transpiration. Figure 3-6 illustrates the transpiration process. As much as 99 percent or more of the water absorbed by the roots may be transpired. Although transpired, it is vital to the life processes of the plant. That portion which evaporates has a cooling effect and serves to create a pressure pull within the conductive tissues which helps to explain the movement of solutes and other materials in the plant's vascular system.

External factors such as light, temperature, humidity, wind velocity, and soil moisture influence the rate of transpiration in plants. The stomates open in response to light; thus, transpiration is much greater in the daytime than at night. Response by the stomates to opening and closure is related to changes within

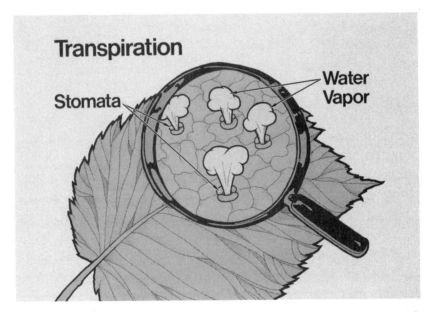

Figure 3-6. An illustration representing the loss of water vapor through stomata in the transpiration process.

the living cells which are affected by external and internal factors.

Modification of plant structure is a natural process which is a result of plant adaptability to climatic factors. Waxy cuticles, thickened layers, decreased leaf surface area, fewer stomates, and hairy leaf surfaces are natural responses which modify transpiration rate and allow for greater resistance to moisture stress.

A quantity factor has been developed to express the ratio of total water absorbed to the total amount of dry matter produced by the plant. This is termed the "transpiration ratio."

The water requirement of a high yielding, properly fertilized crop is essentially the same as a low yielding, poorly fertilized crop. High yields reduce the water required per unit of production and result in better water use efficiency.

# FACTORS AFFECTING GROWTH

How fast a plant grows and the shape or form it assumes are determined by internal and external factors. Heredity is the predominant internal factor, and environment is the major external factor.

## HEREDITY

The tendency for an offspring to display the characteristics of its parents is known as heredity. If we consider a wheat plant — the length and strength of the stem; the shape and texture of the leaf; the number of spikelets in the head; the shape, color, and surface of the glumes; the weight of the kernel; the character of the grain; the yield of the seed; and the resistance to cold, drought, and disease, together with many other characteristics, have all been shown to be inheritable. The male and female sexual cells (gametes) which unite to produce the offspring contain these specific characteristics, and the offspring is a product of that union.

The genes are the remarkable minute pieces of protoplasm located on the chromosomes of the cells that carry the genetic characteristics of the organism. These genes provide the map or the blueprint for all developing cells. Whether the plant is tall or short, the leaf is round or pointed, the flower is red or white, in short, the very genetic makeup of the plant is established by the pattern set by the combination of genes.

Mendel is known as the father of plant genetics. He set down the theory of inheritance and demonstrated that theory. Using the garden pea he was able to single out a particular characteristic, such as flower color or smoothness of the seed, and demonstrate how this characteristic was inherited. He kept accurate records of the crosses he made, knew the ancestry of each individual, and found the number of individuals in the offspring which had the characteristics of the parents. Although the science of genetics is much more complex today, the painstaking and careful scientific methods established by Mendel are still followed in genetic experimentation.

## GROWTH VS. TIME

The growth of a cell, an organ, or a whole plant does not proceed at a uniform rate. Growth starts slowly, gradually increases until a maximum rate is reached and then slows until it ceases altogether. A characteristically S-shaped curve is obtained if total growth is plotted against time. This is illustrated in Figure 3-7.

Fluctuations in temperature, moisture supply, or other environmental conditions may cause irregularities in the curve, but if the whole period of growth is considered, the shape of the curve will remain the same. When nutrient uptake is plotted against time, the accumulation of nutrients closely follows the growth curve. Note that it precedes the growth since the nutrients must be present for growth to occur. A temporary shortage of

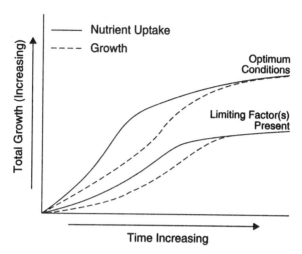

Figure 3-7. Characteristic nutrient uptake and growth curve.

nutrients will cause irregularities in the curve, and a severe shortage will essentially stop growth.

## TEMPERATURE

Temperature has a marked effect upon plant growth and is one of the most important factors determining the distribution of plants over the earth's surface. The temperatures within which plants are able to grow are often designated as minimum, optimum, and maximum. The points below and above which growth ceases are designated minimum and maximum, while the temperature at which growth proceeds best is termed optimum. These points are not fixed during the life of the plant nor are they the same for all parts of the plant. In general, for temperate climates, the minimum falls somewhere above freezing, optimum at about 80° to 90°F, and maximum around 110° to 120°F.

The direct effect of temperature must be evaluated in terms of its effect on such basic processes as photosynthesis, respiration, water and nutrient absorption, chemical processes within the plant, etc. Also, temperature has an effect upon the soil and chemical transformations within the soil. These are intimately related to root activity as illustrated by reduced uptake of phosphorus from a cool soil. Microbial activity is accelerated as temperature of the soil reaches optimum and results in the release of nutrients from soil organic matter and plant residues.

## RADIANT ENERGY

The amount, quality, and duration of sunlight play an important part in plant growth and development. Most plants are able to reach maximum growth at less than full sun intensity; however, this is often modified by density of plant canopy and shading. Some plants are better able to use the maximum sunlight. Genetic characteristics modify the growth rate.

The quality of light directly affects plant growth; for example, certain wavelengths of light trigger germination. Generally, however, light quality is not controllable in most agricultural situations. Except for special situations, it is quite well established that the full spectrum of sunlight is generally most satisfactory for plant growth.

## WATER AND GROWTH

The importance of water to plant growth is readily apparent. Water is required in photosynthesis, is a part of protoplasm, and serves as a vehicle for translocation of food and mineral elements. The availability of water may influence the form, structure, and nature of plant growth. Plants absorb more water than any other soil constituent. As previously indicated, a large part of the moisture taken up is transpired. Since water is such an important factor in plant growth and overall crop production, a separate chapter, Chapter 2, is devoted to this subject.

## THE ATMOSPHERE AND GROWTH

Quantities of nitrogen, oxygen, and carbon dioxide do not vary except locally in the atmosphere. Air normally contains 78 percent nitrogen, 21 percent oxygen, and 0.03 percent carbon dioxide. No higher plant is known which can make direct use of elemental nitrogen except through the action of certain microorganisms. All plants need oxygen for respiration, and the atmosphere is the most common source. Some plants are able to utilize the oxygen from oxidized compounds, such as nitrates and sulfates, but this is normally confined to microorganisms or specialized plants. All photosynthesizing plants require carbon dioxide, and this becomes the basic process by which carbon is fixed into organic matter.

The atmosphere often contains gases, particulate matter, and other contaminants which can have a direct effect on plant growth. Some of these effects are positive. For example, sulfur dioxide at low levels can be absorbed by the aerial portions of some plants, and much of the nutrient need for sulfur can be satisfied this way.

Specific damage to growing plants has been observed from the pollutants in air. Such products as ozone, PAN (peroxyacetyl nitrates), excess sulfur dioxide, ethylene, fluorides, and others come from motor vehicles, combustion of fuels, organic solvents, and other sources. Most damage has been noted on fruit and vegetable crops, pine trees, and flowers. Usually the injury shows up on leaves, but sometimes the plants are stunted or produce poorly.

## MINERAL NUTRIENT REQUIREMENTS AND GROWTH

Present information indicates the need for 16 elements in plant growth. These are carbon, hydrogen, oxygen, nitrogen, phosphorus, potassium, calcium, magnesium, sulfur, boron, chlorine, copper, iron, manganese, molybdenum, and zinc. Many more elements are found in plants but their essentiality has not been established. Some of these, listed alphabetically, are as follows: aluminum, arsenic, barium, bromine, cobalt, fluorine, iodine, lithium, nickel, selenium, silicon, sodium, strontium, titanium and vanadium. This is not a complete list since practically all of the known elements have been isolated at one time or another from plant materials. Functions of nutrient elements in the plant, their supply in the soil, and additions in fertilizers are discussed in other chapters.

By necessity, many of the factors associated with plant growth have been omitted from this short chapter. The discussion has been simplified, and much of the basic biology, chemistry, and related scientific disciplines has only been touched upon. It was written to help the student, the farmer, or anyone interested to understand something about plant growth.

Anyone desiring to become better acquainted with the subject of plant growth is referred to the Supplementary Reading at the end of this chapter.

## PHOTOPERIODISM

Photoperiodism is a term describing the behavior of a plant in relation to day length. Based upon its reaction to day length, a plant is classified as short-day, long-day, or indeterminate. Short-day plants flower only under short-day conditions. If they are grown under long-day photoperiods, they will not flower and will continue to grow vegetatively. Examples of short-day plants are most of the spring flowers and such autumn-flowering plants as ragweed, asters, and scarlet sage.

Long-day plants attain their flowering stage only when the length of day falls within certain limits, usually 12 hours or longer. Such plants are the radish, lettuce, grains, clover, and many others that normally bloom in midsummer.

Still other plants such as the tomato, cotton, and buckwheat complete their reproductive cycles over a wide range of day lengths. These plants are termed indeterminate.

Length of day has also been reported to have an influence on the formation of tubers and bulbs; the character and extent of branching; root growth; abscission or the dropping of leaves, flowers, and other plant parts; dormancy; and other effects.

Photoperiodism plays a large part in whether varieties and species may be adaptable to other areas such as moving a plant species from the north to the south and vice versa. Artificial systems can be set up to induce plants to respond contrary to what they would do if left to grow in the open. The poinsettia is forced to change its growth pattern to produce the colorful display at Christmas by altering its photoperiod.

In reality, the length of the dark period is the controlling factor in photoperiodicity, not the day length. This has been established with experimentation; however, since in normal situations long days have short nights and vice versa, it is natural to assign day length to photoperiodism.

## SUPPLEMENTARY READING

1. *Advanced Plant Physiology.* Malcolm B. Wilkins, ed. John Wiley & Sons. 1984.

2. *Botany: A Brief Introduction to Plant Biology,* Second Edition. T. L. Rost, M. G. Barbour, R. M. Thornton, T. E. Weir and C. R. Stocking. John Wiley & Sons, Inc. 1984.

3. *The Genetic Basis of Plant Physiological Processes.* John King. Oxford University Press. 1991.

4. *Limitations to Efficient Water Use in Crop Production.* H. M. Taylor, W. R. Jordan and T. R. Sinclair, eds. American Society of Agronomy. 1983.

5. *Modern Plant Biology.* H. J. Dittmer. Van Nostrand Reinhold Company. 1972.

6. *Physiological Basis of Crop Growth and Development.* M. B. Tesar, ed. American Society of Agronomy. 1984.

7. *Physiology of Plant Growth and Development.* M. B. Wilkins. McGraw-Hill Book Company. 1969.

8. *Plant Growth.* M. Black and J. Edelman. Harvard University Press. 1970.

9. *Plant Physiology,* Second Edition. F. B. Salisbury and C. W. Ross. Wadsworth Publishing Co. 1979.

10. *Soil Conditions and Plant Growth,* Tenth Edition. E. W. Russell. Longman, Inc. 1974.

# Chapter 4

# *Essential*
# *Plant Nutrients*

There are more than 100 chemical elements known today. Only 16 of these have been shown to be essential to plants (see Figure 4-1). Others may be found to be essential in the future. A few have already demonstrated an ability to stimulate plant growth under certain conditions.

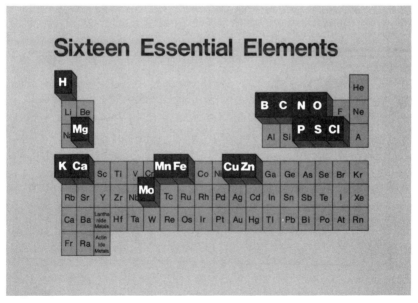

Figure 4-1. Periodic table of elements highlighting the 16 essential plant nutrients.

Three of the 16 essential elements, carbon, hydrogen, and oxygen, are taken primarily from air and water. The other 13 are normally absorbed from soil by plant roots. These 13 elements are divided into three groups: primary nutrients, secondary nutrients, and micronutrients. This grouping separates the elements on the basis of relative amounts required for plant growth. All of these elements are equally essential, regardless of amounts required, and nothing can be substituted for them.

## CARBON, HYDROGEN, AND OXYGEN

Carbon forms the skeleton for all organic molecules. Hence it is a basic building block for plant life. Carbon is taken from the atmosphere by plants in the form of carbon dioxide. Through the process of photosynthesis, carbon is combined with hydrogen and oxygen to form carbohydrates (see page 77). Further chemical combinations, some with other essential elements, produce the numerous substances required for plant growth.

Oxygen is required for respiration in plant cells whereby energy is derived from the breakdown of carbohydrates. Many compounds required for plant growth processes contain oxygen. Hydrogen, along with oxygen, forms water, which constitutes a large proportion of the total weight of plants. Water is required for transport of minerals and plant food, and it also enters into many chemical reactions necessary for plant growth. Hydrogen is also a constituent of many other compounds necessary for plant growth. Since carbon, hydrogen, and oxygen are supplied to plants primarily from the air and water, the concern over their supply is somewhat different from that of the other 13 essential elements.

# *PRIMARY*
# *PLANT NUTRIENTS*

## *NITROGEN*

Nitrogen is taken up by plants primarily as nitrate ($NO_3^-$) or ammonium ($NH_4^+$) ions. Plants can utilize both of these forms of nitrogen in their growth processes.

Most of the nitrogen taken up by plants is in the nitrate form. There are two basic reasons for this. First, nitrate nitrogen is mobile in the soil and moves with soil water to plant roots where uptake can occur. Ammonic nitrogen, on the other hand, is bound to the surfaces of soil particles and cannot move to the roots. Second, all forms of nitrogen fertilizer added to soils are changed to nitrate under proper conditions of temperature, aeration, moisture, etc., by soil organisms.

Nitrogen is utilized by plants to synthesize amino acids which in turn form proteins. The protoplasm of all living cells contains protein. Nitrogen is also required by plants for other vital compounds such as chlorophyll, nucleic acids, and enzymes.

Symptoms of nitrogen deficiency in plants include:

1. Slow growth; stunted plants.

2. Yellow-green color (chlorosis).

3. "Firing" of tips and margins of leaves beginning with more mature leaves.

Chlorosis is usually more pronounced in older tissue since nitrogen is mobile within plants and tends to move from older to younger tissue when nitrogen is in short supply.

## SOIL NITROGEN

Most of the nitrogen in soils is unavailable to growing crops because it is tied up in organic matter. Only about two percent of this nitrogen is made available to crops each year. Since western soils generally contain relatively small amounts of organic matter, the amount of nitrogen made available for crop use each year is not great, perhaps about 20 pounds per acre.

Many reactions involving nitrogen occur in the soil. Most of them are the result of microbial activity. Nitrogen is made available to crops from organic matter through two of these reactions. Protein and allied compounds are broken down into amino acids through a reaction called *aminization*. Soil organisms acquire energy from this digestion. They also utilize some of the amino nitrogen in their own cell structure. Ammonic nitrogen is formed by the second reaction, which converts amino compounds into ammonia ($NH_3$) and ammonium ($NH_4^+$) compounds. This reaction is called *ammonification*. The two reactions, aminization and ammonification, are referred to as *mineralization*.

Ammonic forms of nitrogen are changed to nitrate by two distinct groups of bacteria. *Nitrosomonas* and *Nitrosococcus* convert ammonia to nitrite:

$$2NH_4^+ + 3O_2 \longrightarrow 2NO_2^- + 2H_2O + 4H^+ + energy$$

ammonium   oxygen             nitrite    water    hydrogen
nitrogen                                           ions

*Nitrobacter* oxidizes nitrite to nitrate:

$$2NO_2^- + O_2 \longrightarrow 2NO_3^- + energy$$

nitrite    oxygen                  nitrate

This two-step reaction is called *nitrification*. The reactions occur readily under conditions of warm temperature, adequate oxygen and moisture, and optimum pH.

At 75°F, nitrification may be completed in one to two weeks

(Figure 4-2); at 50°F, 12 weeks or more may be required; and much longer at cooler temperatures.

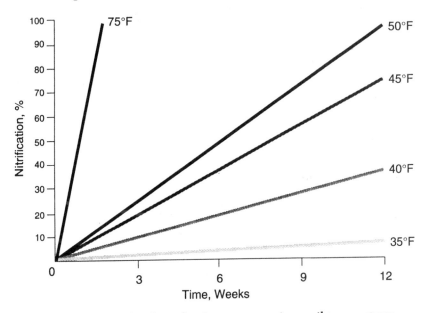

Figure 4-2. Generalized nitrification rates at various soil temperatures.

Nitrogen may be lost from the soil to the atmosphere by reactions that convert nitrate to gaseous compounds of nitrogen. This process is called *denitrification*. Under anaerobic conditions caused by excessive moisture and/or soil compaction, certain bacteria are capable of removing oxygen from chemical compounds in the soil to meet the needs of their life processes. When nitrate is used, various gases such as nitrous oxide ($N_2O$), nitric oxide (NO) and nitrogen ($N_2$) are formed. The reaction can be represented as follows:

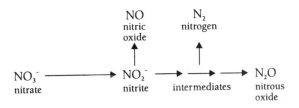

As these gases are lost from the soil into the atmosphere, there is a loss of crop-producing nitrogen from the soil.

## THE NITROGEN CYCLE

The atmosphere contains approximately 78 percent nitrogen. It is estimated that over every acre of land there are some 35,000 tons of nitrogen. In order for crops to utilize this nitrogen, it must be combined with hydrogen or oxygen. This process is called *nitrogen fixation*. Nitrogen may be fixed by various soil organisms. Some of these live in nodules on roots of legumes, and others are free living organisms. Lightning also fixes smaller amounts of nitrogen which are carried into the soil by rain. The fertilizer industry fixes several million tons of nitrogen each year in various nitrogen fertilizers.

Figure 4-3 shows that nitrogen fixation by these various means provides nitrogen for growing crops. Crops in turn provide nitrogen for animals which use the crops for food. Plant

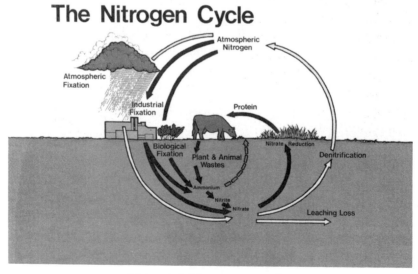

Figure 4-3. The nitrogen cycle.

plants to form nucleic acids (DNA and RNA) and many other vital compounds. It is used in storage and transfer of energy through energy-rich linkages (ATP and ADP) which will be used for growth and reproduction.

Phosphorus stimulates early growth and root formation. It hastens maturity and promotes seed production. Phosphorus supplementation is required most by crops under the following circumstances: (1) growth in cold weather, (2) limited root growth, and (3) fast top growth. Lettuce is an example of a very highly responsive crop due to the conditions outlined. Legumes, such as alfalfa and beans, are large users of phosphate fertilizers. Least responsive crops are trees and vines with extensive root systems raised in warm climates with long summers.

Symptoms of phosphorus deficiency in plants include:

1. Slow growth; stunted plants.

2. Purplish coloration on foliage of some plants (older leaves first).

3. Dark green coloration with tips of leaves dying.

4. Delayed maturity.

5. Poor grain, fruit or seed development.

## POTASSIUM

Potassium is taken up by plants in the form of potassium ions ($K^+$). It is not synthesized into compounds, such as occurs with nitrogen and phosphorus, but remains in ionic form within cells and tissues. Potassium is essential for translocation of sugars and for starch formation. It is required in the opening and closing of stomata by guard cells which is important for efficient water use. Potassium encourages root growth and increases crop resistance to disease. It produces larger, more uniformly distributed xylem vessels throughout the root system. Potassium increases size and quality of fruits, grains, and vegetables and is essential for high-quality forage crops. It also increases winter hardiness of perennials.

Soils may contain 40,000 to 60,000 pounds of potassium per acre. About 90 to 98 percent of the potassium occurs in primary minerals and is unavailable to crops. From 1 to 10 percent is trapped in expanding lattice clays and is only slowly available. Between 1 and 2 percent is contained in the soil solution and on exchange sites and is readily available to crops. Figure 4-5 illustrates the various forms of potassium in soil.

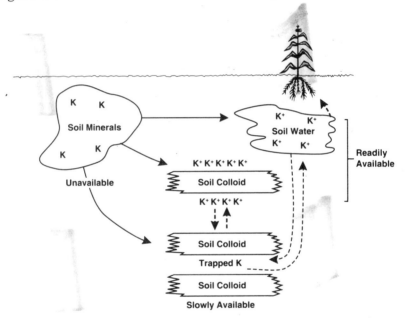

Figure 4-5. Availability of soil potassium.

Potassium has been found to be most required by crops with a very high carbohydrate production. The most responsive vegetable has been the potato, which again is a high producer of carbohydrate as starch in the tubers. Potassium is removed from the soil in larger amounts where the entire vegetative growth is removed such as in high yields of silage, hays, and celery (see Table 4-1).

Symptoms of potassium deficiency in plants include:

1. Tip and marginal "burn" starting on more mature leaves.

2. Weak stalks; plants "lodge" easily.

3. Small fruit or shriveled seeds.

4. Slow growth.

**Table 4–1**
**Plant Food Utilization by Various Crops[1]**

| Crop | Yield | Pounds per Acre | | |
|---|---|---|---|---|
| | | N | $P_2O_5$ | $K_2O$ |
| *Field crops* | | | | |
| Barley | 2½ t. (104 bu.) | 160 | 60 | 160 |
| Canola (whole plant) | 4,000 lbs. (80 bu.) | 240 | 120 | 190 |
| Corn (grain) | 5 t. (179 bu.) | 240 | 100 | 240 |
| Corn (silage) | 30 t. | 250 | 105 | 250 |
| Cotton (lint) | 1,500 lbs. | 180 | 65 | 125 |
| Grain sorghum | 4 t. (143 bu.) | 250 | 90 | 200 |
| Oats | 3,200 lbs. (100 bu.) | 115 | 40 | 145 |
| Rice | 7,000 lbs. | 110 | 60 | 150 |
| Safflower | 4,000 lbs. | 200 | 50 | 150 |
| Sugar beets | 30 t. | 255 | 60 | 550 |
| Wheat | 3 t. (100 bu.) | 175 | 70 | 200 |
| *Vegetable crops* | | | | |
| Asparagus | 3,000 lbs. | 95 | 50 | 120 |
| Beans (snap) | 10,000 lbs. | 175 | 40 | 200 |
| Broccoli | 18,000 lbs. | 80 | 30 | 75 |
| Cabbage | 35 t. | 270 | 65 | 250 |
| Celery | 75 t. | 280 | 165 | 750 |
| Lettuce | 20 t. | 95 | 30 | 200 |
| Potatoes (Irish) | 500 cwt. | 270 | 100 | 550 |
| Squash | 10 t. | 85 | 20 | 120 |
| Sweet potatoes | 15 t. | 155 | 70 | 315 |
| Tomatoes | 30 t. | 180 | 50 | 340 |

*(Continued)*

## Table 4–1 (Continued)

| Crop | Yield | Pounds per Acre | | |
|---|---|---|---|---|
| | | N | P$_2$O$_5$ | K$_2$O |
| *Fruit and nut crops* | | | | |
| Almonds (in shell) | 3,000 lbs. | 200 | 75 | 250 |
| Apples | 15 t. | 120 | 55 | 215 |
| Cantaloupes | 30 t. | 220 | 70 | 400 |
| Grapes | 15 t. | 125 | 45 | 195 |
| Oranges | 30 t. | 265 | 55 | 330 |
| Peaches | 15 t. | 95 | 40 | 120 |
| Pears | 15 t. | 85 | 25 | 95 |
| Prunes | 15 t. | 90 | 30 | 130 |
| *Forage crops* | | | | |
| Alfalfa | 8 t. | 480 | 95 | 480 |
| Bromegrass | 5 t. | 220 | 65 | 315 |
| Clover-grass | 6 t. | 300 | 90 | 360 |
| Orchardgrass | 6 t. | 300 | 100 | 375 |
| Sorghum-sudan | 8 t. | 325 | 125 | 475 |
| Timothy | 4 t. | 150 | 55 | 250 |
| Vetch | 7 t. | 390 | 105 | 320 |

[1]Total uptake in harvested portion.

# SECONDARY PLANT NUTRIENTS

## CALCIUM

Calcium is absorbed by plants as the calcium ion (Ca$^{++}$). It is an essential part of cell wall structure and must be present for the formation of new cells. It is believed to counteract toxic effects of oxalic acid by forming calcium oxalate in the vacuoles of cells. Calcium is non-mobile in plants. Young tissue is affected first under conditions of deficiency.

Calcium has been required as a foliar spray for certain

Plate 4-1. Nitrogen deficiency on corn. Lower leaves show characteristic yellowing (chlorosis), while upper portion of plants remains green.

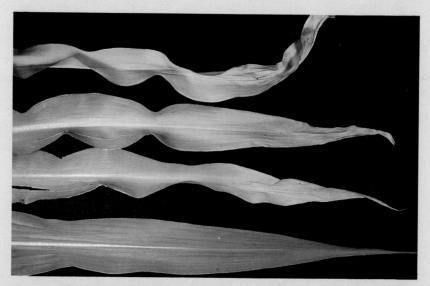

Plate 4-2. Nitrogen deficiency on corn leaves. Lower leaf is normal, upper leaves show characteristic yellowing and "firing" of tissue. Yellowing starts at tips of leaves and moves toward the base in a "V" pattern down the mid-rib.

Plate 4-3. Nitrogen deficiency on sugar beets. Plants become generally light green. Chlorosis is uniform over individual leaves, but is more intense on the older leaves which tend to lie flat on the ground and eventually die.

Plate 4-4. Phosphorus deficiency on corn. Stunted growth and purple coloration are characteristic symptoms on many plants. Symptoms usually appear during early growth, while soils are cold and root systems are small.

Plate 4-5. Phosphorus deficiency on potatoes. Plants appear stunted and are darker green than normal. As the severity of the deficiency increases, the leaves roll upward which exposes the gray-green lower surface, giving the field a normal green color again.

Plate 4-6. Potassium deficiency on alfalfa—two different symptoms. Typically spots develop on margins of more mature leaves and may coalesce to produce entire yellow margins. A second symptom is a white chlorosis at the margins and toward the leaf tips associated with high sodium in the tissues. The two symptoms do not appear on the same plant.

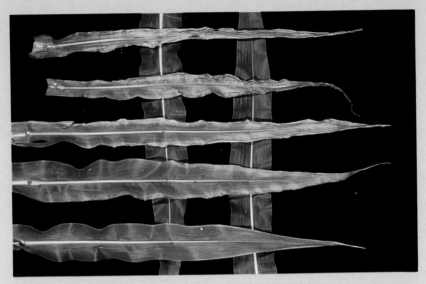

Plate 4-7.  Potassium deficiency on corn.  Leaves show characteristic potassium deficiency symptoms with chlorotic (yellow) and necrotic (dead) tissue along margins of more mature leaves.

Plate 4-8.  Potassium deficiency on cotton.  Leaves develop interveinal chlorosis, thicken, curl, and become progressively more "bronzed" between the veins and along the margins.  Older leaves are affected first in early season.  However, in mid-season during boll formation, symptoms appear on young leaves near stem terminals.

Plate 4-9. Potassium deficiency on grapes. Symptoms typically appear in early summer on the middle portion of the shoots. Leaves become chlorotic beginning at the margins and progressing inward between the veins. Leaves tend to cup downward. In white wine varieties (white variety Chardonnay shown here) leaves become yellow or yellow-bronze. Red varieties show stronger bronzing and a reddening.

Plate 4-10. Potassium deficiency on potatoes. Young, fully expanded leaves become crinkly and develop a glossy sheen with a slightly black pigmentation. Marginal leaf scorching develops followed by necrosis and browning.

Plate 4-11. Sulfur deficiency on alfalfa. Symptoms include stunted growth and general yellowing of foliage. Young leaves are usually affected first.

Plate 4-12. Sulfur deficiency on corn. Symptoms are most often seen on young plants. These include interveinal yellowing (chlorosis) and stunted growth.

Plate 4-13. Magnesium deficiency on citrus (orange). Chlorosis of tip and margins of more mature leaves produces a "Christmas tree" pattern along mid-rib of leaf.

Plate 4-14. Blossom end rot on tomatoes (calcium deficiency). Terminal buds die and fruit breaks down at blossom ends. This can occur even in calcareous soil due to inadequate translocation of calcium within the plant for rapid growing and high yielding varieties.

Plate 4-15. Boron deficiency on alfalfa. Yellowing and reddening develop on the upper leaves. The internodes of the top growth become progressively shorter, resulting in a "rosette" appearance. Growing points become dormant or die.

Plate 4-16. Boron deficiency on apple. Reproduction is affected including fewer flowers produced, lower flower retention, lower pollen germination, and slower pollen tube growth. Vegetative symptoms include thickened, brittle leaves and rosetting and dieback of the growing points.

Plate 4-17. Iron and manganese deficiencies on citrus (lemon). The two center leaves show typical iron deficiency symptoms: green veins with yellowing of interveinal areas of young leaves. The two outer leaves show characteristic manganese deficiency. Younger leaves show interveinal chlorosis with gradation of pale green coloration, with darker color next to the veins.

Plate 4-18. Iron deficiency on milo (grain sorghum). Early symptoms include interveinal chlorosis of young leaves. As deficiency progresses, entire leaves turn yellow, then almost white. Usually symptoms appear in discrete areas of the field commonly associated with high soil pH.

Plate 4-19. Iron deficiency on cherry. Interveinal chlorosis develops on young leaves with veins showing as a fine network of green on a background of yellow. As deficiency progresses, leaves become entirely yellow. Symptoms often appear on poorly aerated soils.

Plate 4-20. Zinc deficiency on citrus (orange). Small, narrow leaves have yellow mottling between the veins. This deficiency affects younger leaves first.

Plate 4-21. Zinc deficiency on cotton. Interveinal chlorosis starting on younger leaves. Affected leaves are smaller than normal, and as the deficiency progresses, they become thick and brittle.

Plate 4-22. Zinc deficiency on kidney beans. Interveinal chlorosis of younger leaves exhibiting a mottle appearance.

Plate 4-23. Boron toxicity on tomatoes. Chlorosis and necrosis develop on margins of leaves. Toxicities occasionally occur where boron content of soil or irrigation water is inherently high.

Plate 4-24. Chloride toxicity on walnut. Marginal necrosis develops on affected leaves. Tissue analysis may be required to differentiate chloride toxicity from general salt damage or wind burn.

celery varieties to prevent a disorder of the stalk called "brown checking." It is also used in the fruit industry to prevent certain fruit disorders, such as bitter pit of apples. Calcium is so generally abundant that its only other requirement as a fertilizer nutrient has been on very acid soils where lime is required.

Symptoms of calcium deficiency in plants include:

1. Death of growing points (terminal buds) on plants. Root tips also affected.

2. Abnormal dark green appearance of foliage.

3. Premature shedding of blossoms and buds.

4. Weakened stems.

5. Fruit disorders.

## MAGNESIUM

Plant uptake of magnesium is in the form of the magnesium ion ($Mg^{++}$). The chlorophyll molecule contains magnesium. It is therefore essential for photosynthesis. Magnesium serves as an activator for many plant enzymes required in growth processes.

Magnesium is mobile within plants and can be readily translocated from older to younger tissue under conditions of deficiency.

The availability of magnesium is generally high in western soils, but is more often deficient than calcium. The most common use of magnesium fertilizer has been on celery and citrus. Crops such as these likely need magnesium to balance the generally high use of potassium from fertilizers and manure. Crops growing in sandy soils may also show deficiencies.

Symptoms of magnesium deficiency in plants include:

1. Interveinal chlorosis (yellowing) in older leaves.

2. Curling of leaves upward along margins.

3. Marginal yellowing with green "Christmas tree" area along mid-rib of leaf.

## SULFUR

Uptake of sulfur by plants is in the form of sulfate ions ($SO_4^-$). Sulfur may also be absorbed from the air through leaves in areas where the atmosphere has been enriched with sulfur compounds from industrial sources.

Sulfur is a constituent of three amino acids (cystine, methionine, and cysteine) and is therefore essential for protein synthesis. It is essential for nodule formation on legume roots. Sulfur is present in oil compounds responsible for the characteristic odors of plants such as garlic and onion.

Sulfur deficiencies are widespread throughout the United States and have been identified under a wide range of soil and climatic conditions. The sulfate ion is leachable similar to nitrate, but to a lesser extent. In low rainfall areas it is frequently found precipitated in the soil profile as gypsum ($CaSO_4$). Deficiencies have become more common in recent years due to a number of factors including higher air quality standards relative to emissions and higher crop yields. This has prompted some to call sulfur the "fourth major nutrient."

Symptoms of sulfur deficiency in plants include:

1. Young leaves light green to yellowish color. In some plants, older tissue may be affected also.

2. Small and spindly plants.

3. Retarded growth rate and delayed maturity.

# MICRONUTRIENTS

Even though micronutrients are used by plants in very small amounts, they are just as essential for plant growth as the larger amounts of primary and secondary nutrients. Care must be exercised in the use of micronutrients, since the difference between deficient and toxic levels often is small. Micronutrients should not be applied as a "shotgun" application to cover possible deficiencies. They should be applied only when the need has been demonstrated.

## ZINC

Zinc is an essential constituent of several important enzyme systems in plants. It controls the synthesis of indoleacetic acid, an important plant growth regulator. Terminal growth areas are affected first when zinc is deficient. Zinc is absorbed by plants as the zinc ion ($Zn^{++}$).

Zinc has been the micronutrient most often needed by western crops. Citrus generally is given zinc as part of a foliar spray program one to several times a year. Many other tree crops, grapes, beans, onions, tomatoes, cotton, rice, and corn have generally required zinc fertilization. Deficiency is most common on soils with neutral or alkaline pH that are sandy or have low organic matter content (including cut areas) and that are very high in available phosphorus.

Symptoms of zinc deficiency in plants include:

1. Decrease in stem length and a rosetteing of terminal leaves.
2. Reduced fruit bud formation.
3. Mottled young leaves (interveinal chlorosis).
4. Dieback of twigs after first year.
5. Striping or banding on corn leaves.

## IRON

Iron is required for the formation of chlorophyll in plant

cells. It serves as an activator for biochemical processes such as respiration, photosynthesis, and symbiotic nitrogen fixation. Iron deficiency can be induced by high levels of manganese or high lime content in soils. Iron is taken up by plants as ferrous ions ($Fe^{++}$).

Iron is usually contained in ample amounts in western soils. Where deficiencies do occur, it is likely due to an imbalance (excess zinc or manganese), high pH or poor aeration. Crops most often affected are grasses such as sorghum and corn, certain tree crops, and ornamentals.

Symptoms of iron deficiency in plants include:

1. Interveinal chlorosis of young leaves. Veins remain green except in severe cases.

2. Twig dieback.

3. In severe cases, death of entire limbs or plants.

## MANGANESE

Manganese serves as an activator for enzymes in growth processes. It assists iron in chlorophyll formation. High manganese concentration may induce iron deficiency. Manganese uptake is primarily in the form of the ion ($Mn^{++}$).

Manganese is generally required with zinc in foliar spraying of commercial citrus. Other tree crops may show deficiencies, but otherwise, there is not common recognition of requirements for this element in fertilizer programs in the west.

Symptoms of manganese deficiency in plants include:

1. Interveinal chlorosis of young leaves. Gradation of pale green coloration with darker color next to veins. No sharp distinction between veins and interveinal areas as with iron deficiency.

2. Development of gray specks (oats), interveinal white

streaks (wheat) or interveinal brown spots and streaks (barley).

## COPPER

Copper is an activator of several enzymes in plants. It may play a role in vitamin A production. A deficiency interferes with protein synthesis. Plant uptake is in the form of ions ($Cu^+$, $Cu^{++}$).

Native copper supply has been recognized only rarely as needing supplementation. These few instances have been with tree crops and some crops on organic soils and sands. Copper can be highly toxic at low levels; therefore, its application is not recommended except where the need has been established.

Symptoms of copper deficiency in plants include:

1. Stunted growth.

2. Dieback of terminal shoots in trees.

3. Poor pigmentation.

4. Wilting and eventual death of leaf tips.

5. Formation of gum pockets around central pith in oranges.

## BORON

In western soils, intensive cropping and irrigation have caused boron deficiencies to become more common over a wide range of soil and climatic conditions.

In soils below pH 9, boron is predominately in solution as boric acid, $H_3BO_3$, and is taken up by plants in this form. It functions in plants in differentiation of meristematic cells. When boron is deficient, cells may continue to divide, but structural components are not differentiated. Boron is also involved in regulating metabolism of carbohydrates in plants.

Past research indicates that boron is not mobile and a continuous supply is necessary at all growing points.

More recent research indicates boron applied to leaves is transported through the phloem to buds, flowers, and fruit. A deficiency is first found in the youngest tissue of the plant.

Symptoms of boron deficiency in plants include:

1. Death of terminal growth, causing lateral buds to develop, producing a "witches'-broom" effect.

2. Thickened, curled, wilted, and chlorotic leaves.

3. Soft or necrotic spots in fruit or tubers.

4. Reduced flowering or improper pollination.

Although boron toxicity can occur, it is rare. Toxicities occur most often on inland desert areas associated with boron contaminated water.

## MOLYBDENUM

Molybdenum is taken up by plants as the molybdate ion ($MoO_4^-$). It is required by plants for utilization of nitrogen. Plants cannot transform nitrate nitrogen into amino acids without molybdenum. Legumes cannot fix atmospheric nitrogen symbiotically unless molybdenum is present.

Molybdenum has been found in quantities toxic to livestock due to high natural concentrations in forage growing in some inland desert areas, including San Joaquin Valley and Nevada. As soils from these areas are farmed more, the problem is being reduced. Deficiencies of molybdenum for legumes have been reported in a number of western states including Washington, Oregon, and Idaho. The application of only a few ounces per acre of molybdenum corrects the problem.

Symptoms of molybdenum deficiency in plants include:

1. Stunting and lack of vigor. This is similar to nitrogen

deficiency due to the key role of molybdenum in nitrogen utilization by plants.

2. Marginal scorching and cupping or rolling of leaves.

3. "Whiptail" of cauliflower.

4. Yellow spotting of citrus.

## CHLORINE

Chlorine is absorbed by plants as the chloride ion, $Cl^-$. It is required in photosynthetic reactions in plants. Until the mid-1980's deficiencies of chlorine in the field were believed to be rare. However, chloride containing fertilizers are now specifically being used to increase disease resistance of many crops. Some nutritional benefits have also been reported. Wheat, barley, corn, and potatoes have benefited from chloride applications in numerous states in the West and High Plains.

Symptoms of chlorine deficiency in plants include:

1. Wilting, followed by chlorosis.

2. Excessive branching of lateral roots.

3. Bronzing of leaves.

4. Chlorosis and necrosis in tomatoes and barley.

5. Leaf spot (lesions) in wheat.

## NUTRIENT BALANCE

Balance is important in plant nutrition. An excess of one nutrient can cause reduced uptake of another. An excess of magnesium in the soil, for example, may inhibit potassium uptake by crops. A heavy application of phosphorus may induce a zinc deficiency in soil that is marginal or low in zinc. Excess iron may induce a manganese deficiency, etc. Positive interactions also occur.

Maintaining a balance of nutrients in the soil is an important management objective agronomically, economically, and environmentally. By judicious use of fertilizers, nutrients which are deficient in soil can be applied to growing crops. The objective of fertilizer programs is to supplement the capacity of soils to supply adequate nutrients to growing crops. The agronomic and environmental issues dealing with fertilizer applications will be presented in detail in Chapter 10, Best Management Practices, and Chapter 11, Fertilizers and the Environment.

## DIAGNOSING NUTRIENT NEEDS

This topic is covered more completely in Chapter 8, Soil and Tissue Testing. No attempt will be made to discuss it other than to point out one or two precautions. Visual symptoms of nutrient deficiencies can be a useful tool for diagnosing problems. Assistance should be obtained from a qualified person, since chlorosis (yellowing) and necrosis (death) of tissues can result from problems other than nutrient deficiencies. Toxicity from excessive amounts of certain elements or damage from herbicides, as well as lack of proper aeration in the root zone, can produce yellowing or death of tissue. A trained person can usually distinguish between these various problems where the average person may have difficulty. Soil and/or tissue analyses should be used to verify nutrient deficiencies and to determine their causes prior to initiating a program for correction. Testing programs are best utilized on a regular basis to predict nutrient requirements before the plants become deficient and show symptoms of stress.

## SUPPLEMENTARY READING

1. *Diagnostic Criteria for Plants and Soils.* H. D. Chapman,

ed. University of California, Division of Agricultural Sciences. 1966.

2. *Hunger Signs in Crops,* Third Edition. H. B. Sprague, ed. David McKay Co., Inc. 1964.

3. *Managing Nitrogen for Groundwater Quality and Farm Profitability.* R. F. Follett, D. R. Keeney and R. M. Cruse, ed. American Soc. of Agronomy. 1991.

4. *Micronutrients in Agriculture.* J. J. Mortvedt, ed. Soil Sci. Soc. of America. 1972.

5. *Nitrogen in Crop Production.* R. D. Hauck, ed. American Soc. of Agronomy. 1984.

6. *Potassium in Agriculture.* R. D. Munson, ed. American Soc. of Agronomy. 1985.

7. *The Role of Phosphorus in Agriculture.* F. E. Khasawneh, ed. American Soc. of Agronomy. 1980.

8. *Soil Fertility and Fertilizers,* Fourth Edition. S. L. Tisdale, W. L. Nelson and J. D. Beaton. The Macmillan Company. 1985.

# Chapter 5

# *Fertilizers — A Source of Plant Nutrients*

Soil serves as a storehouse for plant nutrients and normally provides a substantial amount of the crop's nutrient requirements. Under most conditions, however, crop production can be enhanced by proper application of supplemental nutrients. Any material containing one or more of the essential nutrients that is added to the soil or applied to plant foliage for the purpose of supplementing the plant nutrient supply can be called a *fertilizer*.

The earliest fertilizer materials were animal manures, plant and animal residues, ground bones, and potash salts derived from wood ashes. Three major developments in the nineteenth century in Europe were the forerunners of the modern fertilizer industry:

1839 — The discovery of potassium salt deposits in the German states.

1842 — The treatment of ground phosphate rock with sulfuric acid to form superphosphate.

1884 — The development of the theoretical principles for combining hydrogen and atmospheric nitrogen to form ammonia.

# TYPES OF FERTILIZERS

Based upon their primary nutrient content (N, $P_2O_5$, $K_2O$), fertilizers are designated as single nutrient or multinutrient fertilizers. Single nutrient fertilizers are called "materials" or "simple" fertilizers. Multinutrient fertilizers are referred to as "mixed" or "complex" fertilizers.

Fertilizers containing one or more primary nutrients are given a numerical designation consisting of three numbers. This three number designation is called a "grade" and represents, respectively, the weight percent of nitrogen (N), phosphate ($P_2O_5$), and potash ($K_2O$) contained in the fertilizer.

The fertilizer "ratio" is the relative proportion of each of the primary nutrients. For example, a 12-12-12 grade is a 1-1-1 ratio, and a 21-7-14 grade is a 3-1-2 ratio.

A zero in a grade, or ratio designation, indicates that that particular nutrient is not included in the fertilizer. For example, the grade designation for ammonium nitrate is 34-0-0, and the ratio designation for a 20-10-0 grade is 2-1-0.

## NITROGEN FERTILIZERS

### Anhydrous Ammonia (82-0-0)

Nitrogen from the atmosphere is the primary source of all nitrogen used by plants. This inert gas comprises about 78 percent of the earth's atmosphere. In the fertilizer industry, atmospheric nitrogen is chemically fixed to form ammonia, the principal intermediate for most nitrogen fertilizers (Figure 5-1). The production of ammonia involves the energy intensive reaction of nitrogen and hydrogen. In the presence of a catalyst at temperatures ranging from 400° to 500°C and pressures exceeding 2,200 psig, hydrogen and nitrogen combine to form ammonia according to the following equation:

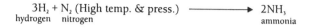

$$3H_2 + N_2 \text{ (High temp. \& press.)} \longrightarrow 2NH_3$$
hydrogen   nitrogen                                                    ammonia

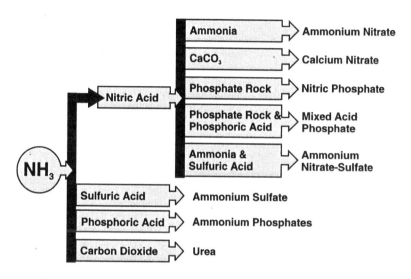

Figure 5-1. Conversions of ammonia to various nitrogen fertilizers.

## Properties of Anhydrous Ammonia

| | |
|---|---|
| Nitrogen content | 82% |
| Color | Colorless |
| Odor | Pungent, sharp |
| Molecular weight | 17 |
| Pounds per gal. at 60°F | 5.14 |
| Pounds per cu. ft. at 60°F | 38.45 |
| Boiling point at 1 atmosphere pressure | –28.03°F |
| Freezing point at 1 atmosphere pressure | –107.86°F |
| Calcium carbonate equivalent (lbs./100 lbs.) | 148 |

Gaseous ammonia is lighter than air. When compressed, it liquifies and is about 60 percent as heavy as water. Ammonia is readily absorbed in water up to concentrations of 30 to 40 percent by weight.

The high vapor pressure of anhydrous ammonia at ordinary temperatures requires that it be transported in pressure containers, generally with a minimum working pressure of 265 psig.

Although the supply of nitrogen from the air is virtually infinite, sources of hydrogen are limited. In the United States, almost all modern ammonia production facilities use natural gas as the hydrogen source. One ton of ammonia requires about 33,000 cubic feet of natural gas to supply the hydrogen required. Alternative sources, such as naphtha, a hydrogen rich hydrocarbon refined from petroleum, are frequently used in foreign plants.

### Aqua Ammonia (20-0-0)

Anhydrous ammonia dissolved in water forms aqua ammonia, commonly called "Aqua." Commercially, it contains 20 percent nitrogen, all in the ammonium form. Under normal temperatures, aqua ammonia has some free ammonia, but the vapor

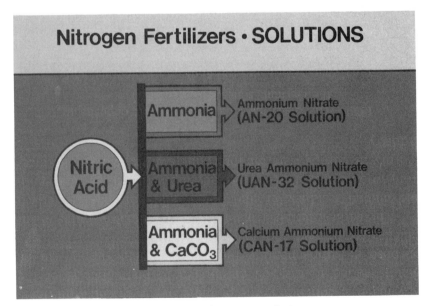

Figure 5-2. Liquid nitrogen solutions.

pressure is low. This makes it possible to store this liquid fertilizer in low-pressure tanks and to apply it with less expensive equipment than required for anhydrous ammonia.

To minimize loss of nitrogen, aqua ammonia, like anhydrous, should be injected below the soil surface or under the water surface when applied in a water run. Because of the much lower free ammonia, direct soil applications of "aqua" need not be injected as deeply as anhydrous ammonia.

### *Ammonium Nitrate (34-0-0)*

Ammonium nitrate was not used extensively as a fertilizer until after World War II. It is manufactured by reacting nitric acid with anhydrous ammonia (Figure 5-2). Nitric acid is produced by the oxidation of $NH_3$ with air in the presence of a catalyst, usually platinum. The initial oxidation reactions are:

$$4NH_3 + 5O_2 \longrightarrow 4NO + 6H_2O$$

ammonia   oxygen                    nitrous oxide   water

$$2NO + O_2 \longrightarrow 2NO_2$$

nitric oxide

The $NO_2$ is then absorbed in water to form nitric acid.

$$3NO_2 + H_2O \longrightarrow 2HNO_3 + NO$$

water                           nitric acid

Nitric acid and ammonia react to form ammonium nitrate.

$$HNO_3 + NH_3 \longrightarrow NH_4NO_3$$

ammonium nitrate

The final product is concentrated, prilled or granulated, and coated to prevent caking. Commercial ammonium nitrate contains 33.5 to 34.0 percent nitrogen.

### Properties of Ammonium Nitrate

| | |
|---|---|
| Nitrogen content | 33.5–34.0% |
| Color | White |
| Molecular weight | 80 |
| Pounds per cu. ft. | 45–62 |
| Angle of repose | 37–40° |
| Critical relative humidity @ 68°F, 20°C | 63.3% |
| Calcium carbonate equivalent (lbs./100 lbs.) | 60.0 |

## Ammonium Nitrate - Lime (26-0-0)

This dry nitrogen fertilizer, widely used in Europe, is a homogeneous mixture containing 40 percent limestone (calcium carbonate) and ammonium nitrate. The powdered lime is added to the concentrated ammonium nitrate solution prior to granulating. Typically, ammonium nitrate-lime contains 26 percent nitrogen, but up to 27.5 percent N is common. In Europe it is referred to as calcium ammonium nitrate (CAN).

## Ammonium Sulfate (21-0-0-24S)

In the Western United States ammonium sulfate is widely used as fertilizer. It contains 21 percent nitrogen and 24 percent sulfur and is one of the oldest forms of solid nitrogen fertilizer.

Directly manufactured ammonium sulfate is made in a neutralizer-crystallizer unit by reacting anhydrous ammonia with sulfuric acid.

The reaction is as follows:

$$2NH_3 + H_2SO_4 \longrightarrow (NH_4)_2SO_4$$

ammonia    sulfuric acid              ammonium sulfate

Most fertilizer grade ammonium sulfate is produced by direct crystallization.

## Properties of Ammonium Sulfate

| | |
|---|---|
| Nitrogen content | 21% |
| Sulfur content | 24% |
| Color | White |
| Molecular weight | 132 |
| Pounds per cu. ft. | 66-68 |
| Angle of repose | 28° |
| Critical relative humidity @ 68°F, 20°C | 81% |
| Calcium carbonate equivalent (lbs./100 lbs.) | 110 |

### Calcium Nitrate (15.5-0-0-19Ca)

The commercial grade of calcium nitrate, chemically known as ammonium calcium nitrate decahydrate, is a white, hygroscopic, granular, water soluble product containing 15.5 percent nitrogen and 19 percent calcium. It is a co-product of the nitric phosphate process wherein nitric acid is reacted with crushed phosphate ore and neutralized with ammonia. The general reaction is:

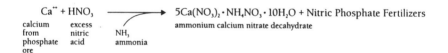

$Ca^{++} + HNO_3 \longrightarrow 5Ca(NO_3)_2 \cdot NH_4NO_3 \cdot 10H_2O$ + Nitric Phosphate Fertilizers

calcium from phosphate ore   excess nitric acid   $NH_3$ ammonia   ammonium calcium nitrate decahydrate

The ammonium calcium nitrate decahydrate is separated from the mother liquor and concentrated via successive evaporation and filtration. The final product is granulated and coated to reduce moisture absorption during storage, handling, and application.

### Nitrate of Soda (16-0-0)

This dry nitrogen material contains 16 percent nitrogen. A very small amount is manufactured in the United States, but a considerable quantity of this mined, natural product is imported from Chile. It is used primarily in the Midsouth and Southeast on specialty crops such as tobacco, fruit, and vegetables as a supplemental nitrogen source. Mined sodium nitrate carries a restricted certification as a fertilizer for organic farming, but the quantity used in this fashion is limited.

### Urea (46-0-0)

Urea contains the highest nitrogen percentage of all the solid fertilizers. It is widely used as a nitrogen fertilizer for plants and as a protein substitute in ruminant animal feeds. In production, ammonia and carbon dioxide are reacted in a special vessel at temperatures between 170° and 210°C and pressures ranging from 170 to 400 atmospheres (2,500 to 6,000 psig). The basic reactions are:

$$2NH_3 + CO_2 + H_2O \longrightarrow (NH_4)_2 CO_3$$

ammonia  carbon  water                     ammonium carbonate
dioxide

$$(NH_4)_2 CO_3 \longrightarrow (NH_2)_2 CO + 2H_2O$$

urea                water

The concentrated liquid from these reactions contains about 80 percent urea. This may be used directly in urea solutions or further concentrated and prilled or granulated to make solid urea. Fertilizer grade urea contains 46 percent nitrogen.

### Properties of Urea

| | |
|---|---|
| Nitrogen content | 46% |
| Color | White |
| Molecular weight | 60 |
| Pounds per cu. ft. | 46-48 |
| Angle of repose | 40° |

Critical relative humidity @ 68°F, 20°C        80.7
Calcium carbonate equivalent (lbs./100 lbs.)    84

Urea is highly water soluble but less corrosive to equipment than many other products. It is a common nitrogen source in dry bulk blends but is incompatible with some fertilizer materials, particularly those containing even small quantities of ammonium nitrate.

## Nitrogen Solutions

Anhydrous ammonia and aqua ammonia have already been discussed. Because of the high water solubility of certain dry fertilizer materials, such as ammonium nitrate and urea, these materials are widely used alone or in combinations in solution. When ammonium nitrate and urea in equal proportions are mixed with water, their individual solubilities are increased. This permits a stable solution containing more nitrogen than is possible with either single component.

Commonly available nitrogen solutions include ammonium nitrate 20 percent (AN-20), urea ammonium nitrate 32 percent (UAN-32), and aqua ammonia (see Table 5-1). Nitrogen solutions may contain ammonium, nitrate and urea nitrogen, the proportion of each being dependent on the components. Such liquids are classified as pressure or non-pressure solutions. Pressure solutions are those with an appreciable vapor pressure because of the presence of more free ammonia than the solution can hold. Thus, they lose nitrogen unless contained in a closed tank.

Like any solution, nitrogen solutions will exhibit the phenomenon of salting out. Salting out is simply the precipitation of the dissolved salts when the temperature drops to a certain point. This point is characteristic of a given solution and is a function of the chemical properties and the content of the components.

Table
Composition and Physical

| | Non-ammoniated Solutions | | | | |
|---|---|---|---|---|---|
| | AN–20 | UREA-20 | UAN | UAN–30 | UAN–32 |
| Total N % | 20 | 20 | 28[1] | 30[1] | 32 |
| NH$_3$ % | | | | | |
| NH$_4$NO$_3$ % | 57.2 | | 39.5 | 42.2 | 44.3 |
| Urea % | | 43.5 | 30.5 | 32.7 | 35.4 |
| Water % | 42.8 | 56.5 | 30 | 25.1 | 20.3 |
| Nitrate N % | 10 | | 7 | 7.4 | 7.8 |
| Ammonia N % | 10 | | 7 | 7.4 | 7.8 |
| Urea N % | | 20 | 14 | 15.2 | 16.4 |
| Spec. grav. @ 60°F | 1.26 | 1.12 | 1.28 | 1.3 | 1.33 |
| Lbs./gal @ 60°F | 10.5 | 9.33 | 10.66 | 10.83 | 11.06 |
| Lbs.N/gal. @ 60°F | 2.1 | 1.87 | 2.98 | 3.25 | 3.54 |
| Vapor pressure psig @ 100°F | | ——— (typically non-pressure) ——— | | | |
| Crystallization temp., °F | 41 | 52 | 1 | 15 | 32 |

[1]Used in colder climates where salting may occur.

[2]Generally sold at 20% N solution since some ammonia loss may occur in handling.

Common in the western states is CAN-17 (17-0-0), a clear solution containing a mixture of calcium nitrate and ammonium nitrate. This liquid fertilizer is produced by acidulation of calcium containing minerals (limestone, phosphate ore) with nitric acid. The acidic calcium nitrate liquor is neutralized with ammonia and the nitrogen content adjusted by the addition of ammonium nitrate. Typically, CAN-17 is used as a supplemental source of nitrogen and calcium for fruit, vegetable, and field crops.

### Properties of CAN-17

| | |
|---|---|
| Nitrogen content | 17% |
| Ammoniacal nitrogen | 5.4-5.8% |
| Nitrate nitrogen | 11.2-11.6% |

## 5–1
## Properties of N Solutions

| Ammoniated Solutions | | | Aqua Ammonia Solutions | |
|---|---|---|---|---|
| 37 | 37 | 41 | 20 | 20.6[2] |
| 16.6 | 15.8 | 22.2 | 24.4 | 25 |
| 66.8 | 58.5 | 65 | | |
| | 7.7 | | | |
| 16.6 | 18 | 12.8 | 75.6 | 75 |
| 11.7 | 10.2 | 11.4 | | |
| 25.3 | 23.2 | 29.6 | 20 | 20.6 |
| | 3.6 | | | |
| 1.19 | 1.17 | 1.14 | 0.912 | 0.911 |
| 9.91 | 9.75 | 9.5 | 7.6 | 7.59 |
| 3.67 | 3.61 | 3.9 | 1.52 | 1.52 |
| 1 | 2 | 10 | 1 | 2 |
| 56 | 28 | 21 | –58 | –103 |

| | |
|---|---|
| Calcium content | 7.6–8.8% |
| Specific gravity (20°C/68°F) | 1.51 |
| Pounds per gal. (20°C/68°F) | 12.6 |
| Vapor pressure | Non-pressure |
| pH | 6.2–6.4 |
| Color | Optional |
| Salting out temperature | 30°F |
| Equivalent acidity (lbs. $CaCO_3$ per 100#N) | 53 |

The corrosive characteristics of nitrogen solutions are presented in Table 5-2. Corrosion inhibitors are added to reduce their chemical attack on carbon (mild) steel often used in tanks and other handling equipment. Ammonium thiocyanate (0.1%)

Table 5–2
Corrosive Characteristics of Non-Pressure
Nitrogen Solutions on Various Metals, Alloys,
and Other Materials

| Not Corroded | Corroded | Materials Destroyed Rapidly |
|---|---|---|
| Aluminum and aluminum alloys, types 3003, 3004, 5052, 5154 and 6061<br>Stainless steel, types 303, 304, 316, 347 and 416<br>Rubber<br>Neoprene<br>Polyethylene<br>Vinyl resins<br>Glass | Carbon steel<br>Cast iron | Copper<br>Brass<br>Bronze<br>Monel<br>Zinc<br>Galvanized metals<br>Usual die castings<br>Concrete |

and borate salts (0.1-0.4 percent) are common inhibitors. Maintaining the solution pH near 7.0 also minimizes corrosion.

## PHOSPHATE FERTILIZERS

The earth's crust in certain areas is richly endowed with natural deposits of fluoroapatite or phosphate rock. Such deposits are the basic source of all phosphorus fertilizer materials. These ores may be of igneous or sedimentary origin, with the latter constituting the bulk of the world's reserves.

The principal world reserves now mined are in North Africa, North America, and the former Soviet Union. The important deposits in the United States are located in four areas: Florida;

the States of Idaho, Montana, Utah and Wyoming; North Caro-
lina; and Tennessee.

The raw ore has a phosphate content of 14 to 35 percent
which must undergo concentration (beneficiation) before final
processing into fertilizer. Beneficiation is a process which involves wet screening, hydroseparation, and concentration by
flotation. The resulting product is then dried and ground, or
slurried prior to shipment to the fertilizer plant. At this stage,
the $P_2O_5$ content is 31 to 33 percent, but most of it is still
unavailable to crops.

There remains the task of converting the concentrated
fluoroapatite ore to more soluble forms of phosphate that can
be utilized by crops. Occasionally, finely ground rock phosphate
is used directly on acid soils as a source of phosphorus, but
it's efficiency is very low.

The pathways used by the industry to produce phosphorus
fertilizer from rock phosphate are shown in Figure 5-3.

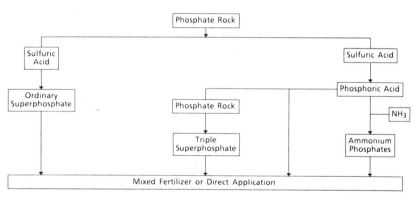

Figure 5-3. Pathways for treating phosphate rock in the production of
phosphatic fertilizers.

### Phosphoric Acid and
### Superphosphoric Acid

Phosphoric acid is an important intermediate in the production of phosphatic fertilizer. There are two methods of producing phosphoric acid: the wet-process and the furnace-grade method. The wet-process method is the principal method used by the fertilizer industry. "Furnace acid," more costly to produce, is used primarily for food and industrial purposes and for highly soluble specialty fertilizer mixes.

Wet-process orthophosphoric acid is produced by the action of sulfuric acid on finely ground phosphate rock. The principal chemical reaction in simplified form is:

$$Ca_{10}F_2(PO_4)_6 + 10H_2SO_4 + 20H_2O \longrightarrow 10CaSO_4 \cdot 2H_2O + 2HF$$

| $Ca_{10}F_2(PO_4)_6$ | $10H_2SO_4$ | $20H_2O$ | $10CaSO_4 \cdot 2H_2O$ | $2HF$ |
|---|---|---|---|---|
| phosphate rock | sulfuric acid | water | gypsum | hydrogen fluoride |

$$+ 6H_3PO_4$$

orthophosphoric acid

The phosphoric acid is separated from the gypsum "cake" by washing and filtration. This raw acid contains about 30 percent $P_2O_5$ and is generally concentrated to the range of 40 to 54 percent $P_2O_5$.

Wet-process acid frequently will be green or black as a result of impurities from compounds of Fe, Al, Ca, Mg, F, and organic matter. These impurities may be present in solution or solid form and some may be beneficial as a source of nutrients.

Several by-products are formed during the manufacture of wet-process acid. The principal one, impure gypsum, represents a major disposal problem for many fertilizer manufacturers.

The trend in the fertilizer industry is to make and use higher analysis compounds. This has prompted the development of superphosphoric acid, a condensation product of orthophosphoric acid. The condensation step is illustrated in Figure 5-4.

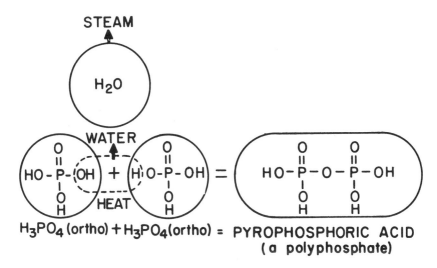

Figure 5-4. Condensation removal of water from orthophosphoric acid to produce pyrophosphoric acid.

The linking of two orthophosphoric acid molecules produces pyrophosphoric acid; three molecules gives tripolyphosphoric acid, and so on, as shown in Figure 5-5.

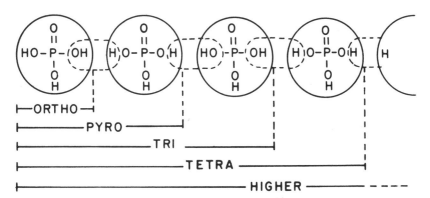

Figure 5-5. The linkage of various members of orthophosphoric acid molecules to produce various polyphosphoric acids.

Collectively, such an acid solution is called polyphosphoric acid or superphosphoric acid and can be defined as a phosphoric acid whose molecular structure contains more than one atom of phosphorus, such as pyrophosphoric acid ($H_4P_2O_7$), tripolyphosphoric acid ($H_5P_3O_{10}$), and tetrapolyphosphoric acid ($H_6P_4O_{13}$). The relationship between the concentration of acid and the type of acid species is shown in Table 5-3.

**Table 5–3**
**Forms of Phosphoric Acid at Various Concentrations[1]**

| Weight % $P_2O_5$ | Equivalent % $H_3PO_4$ | % of Total Phosphorus as | | | | |
|---|---|---|---|---|---|---|
| | | Ortho-phos-phate | Pyro-phos-phate | Tri-poly-phos-phate | Tetra-poly-phos-phate | Higher Poly-phos-phate |
| 54 | 75 | 100 | | | | |
| 68.8 | 95 | 100 | | | | |
| 70 | 97 | 96 | 4 | | | |
| 72 | 99 | 90 | 10 | | | |
| 75.5 | 104 | 53 | 40 | 7 | | |
| 77 | 106 | 40 | 47 | 11 | 2 | |
| 80 | 110 | 13 | 35 | 25 | 14 | 13 |
| 85 | 117 | 2 | 7 | 8 | 11 | 72 |

[1]These data are for furnace-grade acid. Wet-process acid data are generally variable. Differences may occur between wet-process acid manufacturers.

### *Normal Superphosphate (0-20-0)*

Normal superphosphate is produced by reacting sulfuric acid with finely ground phosphate rock. The simplified chemical reaction is:

$$Ca_{10}F_2(PO_4)_6 + 7H_2SO_4 + 17H_2O \longrightarrow 3Ca(H_2PO_4)_2 \cdot H_2O$$

phosphate rock    sulfuric acid    water          monocalcium phosphate monohydrate

$$+ 7CaSO_4 \cdot 2H_2O + 2HF$$

gypsum          hydrogen fluoride

The highly insoluble phosphate in the phosphate rock is converted to monocalcium phosphate monohydrate wherein the phosphate is approximately 85 percent water-soluble. Gypsum, the primary by-product from this reaction, is intimately mixed with the monocalcium phosphate.

In the most common procedure used in the United States, sulfuric acid and phosphate rock are combined and mixed for one to two minutes. The resulting plastic mass is discharged into a compartment called a "den" and retained for 1 to 24 hours until the acidulated phosphate solidifies into a hard block. As it is removed from the den, the block is cut into chunks by mechanical excavators equipped with knives. This coarse material is aired for several weeks in storage to allow completion of the chemical reaction. The cured normal superphosphate is then pulverized and granulated for shipment and usually contains 20 percent $P_2O_5$ and 12 percent sulfur.

### Concentrated Superphosphate (0-45-0)

Similar processes are involved in the production of concentrated superphosphate (sometimes called "triple- or treble-superphosphate"). The principal difference is that concentrated superphosphate is made with phosphoric acid as the acidulant. Finely ground phosphate rock and phosphoric acid in the proper proportions are mixed together. The general reaction is:

$$Ca_{10}F_2(PO_4)_6 + 14H_3PO_4 + 10H_2O \longrightarrow 10Ca(H_2PO_4)_2 \cdot H_2O$$

phosphate    phosphoric  water                       monocalcium phosphate
rock           acid

$$+ \, 2HF$$

hydrogen
fluoride

The end product contains 45 percent $P_2O_5$ and 1 percent sulfur.

Most superphosphates can be granulated through the addition of water and steam in a rotary drum granulator followed by drying and screening.

## NITROGEN-PHOSPHATE COMBINATIONS

### Ammonium Phosphates

The term ammonium phosphate encompasses a wide variety of fertilizers produced by ammoniation of phosphoric acid, often in a mixture with other materials. Such fertilizers may contain ammonium sulfate or ammonium nitrate as an additional source of nitrogen. Ammonium phosphate may be present as the diammonium or monoammonium salt, or a mixture of the two.

Ammonium phosphates are widely used in bulk blends and for direct application. Common ammonium phosphates are monoammonium phosphate 11-52-0 (MAP), diammonium phosphate 18-46-0 (DAP), and ammonium phosphate sulfate 16-20-0-15(S). Liquid forms of ammonium phosphate commonly used are 8-24-0, 9-30-0, and 10-34-0.

Ammonium phosphates are produced by reacting ammonia with phosphoric acid in a preneutralization vessel and further ammoniating the slurry in a rotating ammoniation-granulation unit. Additional nitrogen, phosphate, and sulfuric acid can be introduced into the ammoniation-granulation unit to make the desired grade. The material discharged from the unit is dried, screened, and cooled before being placed in storage.

Solid ammonium polyphosphate fertilizers may be made by ammoniating superphosphoric acid. The acid is ammoniated in a water-cooled reactor at elevated temperature and pressure or in a pipe reactor at high temperature. The product is an anhydrous melt that is granulated by mixing it with solid recycle material in a pugmill followed by drying and cooling. The analysis of the final product is 11-55-0.

### Nitric Phosphates

There are three basic nitric phosphate processes, but all are modifications of the same basic reaction:

$$Ca_{10}F_2(PO_4)_6 + 20HNO_3 \longrightarrow 6H_3PO_4 + 10Ca(NO_3)_2 \cdot 2HF$$

| phosphate rock | nitric acid | | phosphoric acid | calcium nitrate | hydrogen fluoride |

The resulting solution is neutralized with ammonia, concentrated and granulated. Potassium may be added prior to granulation or prilling. Most of the phosphorus is present as high water solubility (>75 percent) MAP and dicalcium phosphate. Modified nitric phosphates are better described as ammonium nitrate phosphates - double salts of ammonium nitrate and ammonium phosphate. The co-product in the process, calcium nitrate liquor, is diverted to neutralization, purification, and granulation as described on page 115 in the production of the commercial grade of calcium nitrate (ammonium calcium nitrate decahydrate).

## POTASSIUM FERTILIZERS

Potassium is found throughout the world in both soluble and insoluble forms. Today only the soluble forms are economically attractive. The largest deposits occur primarily as chlorides and sulfates. Potassium chloride is by far the most important source of potash. Potassium sulfate and potassium nitrate are normally used where the chloride ion may result in poor crop quality or in a buildup of chloride in the soil. Table 5-4 gives the properties of the principal potassium salts.

The greatest volume of potassium produced today is from underground deposits. These are recovered by conventional shaft mining and solution mining in which water is pumped underground to dissolve the ore. The commercial feasibility of a deposit is directly related to its depth, which can vary from a

Figure 5-6. Mining potash ore below the earth's surface.

## Table 5-4
## Properties of Potassium Salts

| | | Potassium Chloride | Potassium Sulfate | Potassium Nitrate | Potassium Thiosulfate |
|---|---|---|---|---|---|
| Color | | white[1] | white | white | clear |
| Molecular weight | | 74.5 | 174 | 101 | 190 |
| Specific gravity | | 1.98 | 2.66 | 2.11 | 1.46 |
| Melting point | | 772°C | 1,067°C | 308°C | — |
| $K_2O$ content, % | | 63 | 54 | 46.6 | 25 |
| Solubility in | 0°C | 13.6 | 3.7 | 5.5 | 25 |
| water (%$K_2O$) | 10°C | 15.0 | 4.6 | 8.1 | 25 |
| | 20°C | 16.1 | 5.4 | 11.2 | 25 |
| | 30°C | 17.1 | 6.2 | 14.6 | 25 |
| | 40°C | 18.1 | 7.0 | 18.2 | 25 |
| | 50°C | 18.9 | 7.7 | 20.5 | 25 |
| | 60°C | 19.8 | 8.6 | 24.2 | 25 |
| | 70°C | 20.6 | 8.9 | 27.0 | — |
| | 80°C | 21.3 | 9.5 | 29.2 | — |
| | 90°C | 22.2 | 10.0 | 31.1 | — |
| | 100°C | 22.9 | 10.5 | 33.2 | — |

[1]Impurities may impart a red color to some fertilizer grades.

few hundred feet to more than 4,000 feet below the surface. Drilling beyond 4,000 feet is cost prohibitive.

The potassium deposits in New Mexico vary from 500 to 2,500 feet below the surface. A deposit in Utah is located at about the 2,700-foot level. The Canadian beds in Saskatchewan tend to be closer to the surface in the north and deeper as they approach the U.S. border. The northern beds now being worked by shaft mining techniques lie at 3,000 to 3,500 feet. At Regina, Saskatchewan, the beds are at about 5,000 feet and at the U.S. border, 7,000 feet below the surface.

An alternative source of potassium is in natural brines in various parts of the world. Currently there are three major natural brine deposits being worked in the world — Searles Lake, California; Great Salt Lake, Utah; and the Dead Sea Works near Sodom, Israel.

Where conventional shaft mining processes are used, the potassium chloride (KCl) is brought to the surface as sylvinite ore. Sylvinite ore is a physical mixture of interlocked crystals of sylvinite (KCl) and halite (NaCl) containing small quantities of dispersed clay and other impurities. The KCl is separated from the NaCl and other minerals by a selective flotation process.

## SECONDARY NUTRIENTS

In the fertilizer industry and in agriculture generally, the importance of the major elements, nitrogen, phosphorus, and potassium, is well established. Secondary and micronutrients are also fully recognized as essential for plant growth. When guarantees are made for secondary and micronutrients, the usual requirement is that they be stated in terms of elemental concentration. Table 5-5 gives the sulfur, calcium, and magnesium contents of most of the commonly used fertilizers.

Calcium sources for crop nutrition are often needed. Soil application of dry or liquid calcium sources and foliar sprays of soluble calcium products are common practices for field

Table
Average Composition of

| Fertilizer Materials | Chemical Formula | Total Nitrogen N % | Available Phosphoric Acid P$_2$O$_5$ % |
|---|---|---|---|
| *Nitrogen materials* | | | |
| Ammonium nitrate | NH$_4$NO$_3$ | 33.5–34 | |
| Monoammonium phosphate | NH$_4$H$_2$PO$_4$ | 11 | 52–55 |
| Ammonium phosphate-sulfate | NH$_4$H$_2$PO$_4$•(NH$_4$)$_2$SO$_4$ | 16 | 20 |
| Ammonium phosphate-nitrate | NH$_4$H$_2$PO$_4$•NH$_4$NO$_3$ | 23–30 | 5–23 |
| Diammonium phosphate | (NH$_4$)$_2$HPO$_4$ | 16–18 | 46–48 |
| Ammonium sulfate | (NH$_4$)$_2$SO$_4$ | 21 | |
| Ammonium thiosulfate | (NH$_4$)$_2$S$_2$O$_3$ | 12 | |
| Anydrous ammonia | NH$_3$ | 82 | |
| Aqua ammonia | NH$_4$OH | 20 | |
| Calcium ammonium nitrate solution | Ca(NO$_3$)$_2$+NH$_4$NO$_3$ | 17 | |
| Calcium nitrate | 5Ca(NO$_3$)$_2$+NH$_4$NO$_3$•10H$_2$O | 15.5 | |
| Sodium nitrate | NaNO$_3$ | 16 | |
| Urea | CO(NH$_2$)$_2$ | 45–46 | |
| Methylene ureas | | 38–41 | |
| Urea ammonium nitrate solution | NH$_4$NO$_3$•CO(NH$_2$)$_2$ | 28–32 | |
| *Phosphate materials* | | | |
| Single superphosphate | Ca(H$_2$PO$_4$)$_2$ | | 18–20 |
| Triple superphosphate | Ca(H$_2$PO$_4$)$_2$ | | 45–46 |
| Phosphoric acid | H$_3$PO$_4$ | | 52–54 |
| Superphosphoric acid | — [2] | | 76–83 |
| *Potash materials* | | | |
| Potassium chloride | KCl | | |
| Potassium nitrate | KNO$_3$ | 13 | |
| Potassium sulfate | K$_2$SO$_4$ | | |
| Potassium thiosulfate | K$_2$S$_2$O$_3$ | | |
| Sulfate of potash-magnesia | K$_2$SO$_4$•2MgSO$_4$ | | |

[1]Equivalent per 100 lbs. of material.
[2]H$_3$PO$_4$, H$_4$P$_2$O$_7$, H$_5$P$_3$O$_{10}$, H$_6$P$_4$O$_{13}$ and other higher phosphate forms.

## 5–5
## Fertilizer Materials

| Water Soluble Potash K$_2$O % | Combined Calcium Ca % | Combined Magnesium Mg % | Combined Sulfur S % | Equivalent Acidity or Basicity in Lbs. CaCO$_3$[1] | |
|---|---|---|---|---|---|
| | | | | Acid | Base |
| | | | | 62 | |
| | | | | 58 | |
| | | | 15 | 88 | |
| | | | 0–7 | 65 | |
| | | | | 70 | |
| | | | 24 | 110 | |
| | | | 26 | 102 | |
| | | | | 148 | |
| | | | | 36 | |
| | 7.6–8.8 | | | 9 | |
| | 19 | | | | 20 |
| | | | | | 29 |
| | | | | 71 | |
| | | | | 60 | |
| | | | | 57 | |
| | 18–21 | | 12 | neutral | |
| | 12–14 | | 1 | neutral | |
| | | | | 110 | |
| | | | | 160 | |
| 60–62 | | | | neutral | |
| 44–46 | | | | | 26 |
| 50–53 | | | 18 | neutral | |
| 25 | | | 17 | 26 | |
| 22 | 0.1 | 11 | 18 | neutral | |

crops, fruits, vegetables, and various horticultural plants. Other sources which provide calcium include soil amendments (lime and gypsum), manure, and irrigation water.

Magnesium sources for crop nutrition include Epsom salts (magnesium sulfate), the double salt of potassium-magnesium sulfate, and magnesium nitrate used for foliar applications on citrus. On acid soils, to which lime is applied, the use of dolomitic lime provides magnesium as well as calcium.

Sulfur for nutrition of crops is provided in many N, P, and K materials. (Consult Table 5-5 for data.) Additional sulfur sources include:

| Source | Percent Sulfur |
|---|---|
| Elemental sulfur | 99 |
| Gypsum | 16–18 |
| Sulfuric acid (95–99%) | 32 |
| Ferrous sulfate | 11.5 |
| Ferric sulfate | 18–19 |
| Calcium polysulfide solution | 23 |
| Ammonium polysulfide solution (20%N) | 40–45 |
| Ammonium bisulfite solution (8.5%N) | 17 |
| Ammonium thiosulfate solution (12%N) | 26 |
| Ammonium sulfate (21%N) | 24 |
| Potassium thiosulfate (25%$K_2O$) | 17 |

Other common sources of sulfur include manure, most river water, rain water, and pesticidal sulfur.

## MICRONUTRIENTS

Properties of the various micronutrient sources vary considerably. A micronutrient material may be completely water-soluble or only slightly soluble. Inorganic sources may be relatively pure compounds or mixtures of compounds containing one or more micronutrients, with or without nonmicronutrient compounds. Organic sources are available as synthetic chelates or natural organic complexes of metal ions. Therefore, one classification of micronutrient sources includes (1) inorganic salts, (2) synthetic chelates, and (3) natural organic complexes.

### Inorganic Salts

Sulfates of Cu, Fe, Mn, and Zn plus borates and molybdates are the most common sources of inorganic micronutrients. The most commonly used boron source is sodium tetraborate. This compound is relatively water-soluble. Soluble boric acid and sodium octaborate (polybor) are often used as foliar sprays.

Water-soluble ammonium and sodium molybdate are the primary sources of molybdenum. Molybdic oxide, a slightly soluble compound, is occasionally used.

See Table 5-6 for composition and properties of inorganic micronutrient materials.

### Table 5-6
#### Inorganic Sources of Micronutrients

| Material | Element | Water Solubility | °F |
|---|---|---|---|
| | (%) | (g/100g $H_2O$) | |
| *Sources of boron* | | | |
| Granular borax—$Na_2B_4O_7 \bullet 10H_2O$ | 11.3 | 2.5 | 33 |
| Sodium tetraborate, anhydrous— $Na_2B_4O_7$ | 21.5 | 1.3 | 32 |
| Solubor®—$Na_2B_8O_{13} \bullet 4H_2O$ | 20.5 | 22 | 86 |
| Ammonium pentaborate— $NH_4B_5O_8 \bullet 4H_2O$ | 19.9 | 7 | 64 |
| *Sources of copper* | | | |
| Copper sulfate—$CuSO_4 \bullet 5H_2O$ | 25.0 | 24 | 32 |
| Cuprous oxide—$Cu_2O$ | 88.8 | i[1] | |

*(Continued)*

## Table 5-6 (Continued)

| Material | Element | Water Solubility | °F |
|---|---|---|---|
| | (%) | (g/100g H₂O) | |
| Sources of copper (continued) | | | |
| Cupric oxide—CuO | 79.8 | i¹ | |
| Cuprous chloride—$Cu_2Cl_2$ | 64.2 | 1.5 | 77 |
| Cupric chloride—$CuCl_2$ | 47.2 | 71 | 32 |
| | | | |
| Sources of iron | | | |
| Ferrous sulfate—$FeSO_4 \bullet 7H_2O$ | 20.1 | 33 | 32 |
| Ferric sulfate—$Fe_2(SO_4)_3 \bullet 9H_2O$ | 19.9 | 440 | 68 |
| Iron oxalate—$Fe_2(C_2O_4)_3$ | 30.0 | very soluble | |
| Ferrous ammonium sulfate— | | | |
| $Fe(NH_4)_2(SO_4)_2 \bullet 6H_2O$ | 14.2 | 18 | 32 |
| Ferric chloride—$FeCl_3$ | 34.4 | 74 | 32 |
| | | | |
| Sources of manganese | | | |
| Manganous sulfate—$MnSO_4 \bullet 4H_2O$ | 24.6 | 105 | 32 |
| Manganous carbonate—$MnCO_3$ | 47.8 | 0.0065 | 77 |
| Manganese oxide—$Mn_3O_4$ | 72.0 | i¹ | |
| Manganous chloride—$MnCl_2$ | 43.7 | 63 | 32 |
| Manganous oxide—MnO | 77.4 | | |
| | | | |
| Sources of molybdenum | | | |
| Sodium molybdate—$Na_2MoO_4 \bullet H_2O$ | 39.7 | 56 | 32 |
| Ammonium molybdate— | | | |
| $(NH_4)_6Mo_7O_{24} \bullet 4H_2O$ | 54.3 | 44 | 77 |
| Molybdic oxide—$MoO_3$ | 66.0 | 0.11 | 64 |
| | | | |
| Sources of zinc | | | |
| Zinc sulfate—$ZnSO_4 \bullet H_2O$ | 36.4 | 89 | 212 |
| Zinc oxide—ZnO | 80.3 | i¹ | |
| Zinc carbonate—$ZnCO_3$ | 52.1 | 0.001 | 60 |
| Zinc chloride—$ZnCl_2$ | 48.0 | 432 | 77 |
| Zinc oxysulfate—$ZnO \bullet ZnSO_4$ | 53.8 | — | — |
| Zinc ammonium sulfate— | | | |
| $ZnSO_4 \bullet (NH_4)_2SO_4 \bullet 6H_2O$ | 16.3 | 9.6 | 32 |
| Zinc nitrate—$Zn(NO_3)_2 \bullet 6H_2O$ | 22.0 | 324 | 68 |

¹An "i" denotes insolubility.

### Synthetic Chelates

A chelating agent is a compound (usually organic) which can combine with a metal ion and form a ring structure between a portion of the chelating agent molecule and the metal. This delays precipitation of the metal ions in the soil. Commercially available synthetic chelating agents and the concentration of micronutrients are shown in Table 5-7.

Table 5–7
Synthetic Chelates

| Chelating Agent | Micronutrient Content, Percent Element[1] | | | |
|---|---|---|---|---|
| | Cu | Fe | Mn | Zn |
| EDTA | 7–13 | 5–14 | 5–12 | 6–14 |
| HEEDTA | 4–9 | 5–9 | 5–9 | 9 |
| NTA | — | 8 | — | 13 |
| DTPA | — | 10 | — | — |
| EDDHA | — | 6 | — | — |

[1]Where a range is shown, the low number indicates the analysis of common liquid forms and the high number indicates the analysis of common dry forms.

Formation constants and chelate-metal stability over pH ranges are important criteria for evaluating different chelating agents. Generally, the stability of metal chelates is greater near neutral than at low or high pH values. This is an important consideration for the formulator incorporating metal chelates into macronutrient fertilizers. If ZnEDTA is mixed with phosphoric acid prior to ammoniation, the complex will break down, but if ZnEDTA is mixed with the ammoniating solution, the complex will remain stable.

### Natural Organic Complexes

Many naturally occurring compounds contain chemically reactive groups similar to synthetic chelating agents. Those used commercially to complex micronutrients are often prepared from by-products of the wood pulp industry. Metal complexes of these compounds have lower stability than the common synthetic chelates. Also, they are more readily broken down by microorganisms in soil. Most are suitable for foliar sprays and fluid fertilizer mixes.

## ORGANIC PRODUCTS

Organic products and organic materials can be classified in several different ways. Strictly speaking, the term *organic* denotes carbon, including that of synthetic origin. However, in popular terminology, organic fertilizers are usually considered naturally occurring compounds. In this category fall the wastes from sewage plants and manures. Generally, they are good soil amendments. They add quantities of organic matter to the soil along with small amounts of plant nutrients. The user should be aware that sewage and industrial wastes may be contaminated with high levels of toxic elements such as cadmium and lead. Continuous use could cause excessive levels of these toxic elements to enter the crops.

Organic material should not be wasted, but treated as a resource. Composts of household residues can be used beneficially in the improvement of the homeowner's garden. The nutrients contained in the compost can be utilized, and the organic matter will help improve soil structure. It is, however, economically unsound to expect commercial farmers to meet their crops' nutrient needs through "organic" products. Organic crop management represents a very small percentage of production agriculture.

## SPECIALTY FERTILIZERS

Considerable effort has been directed toward developing fertilizers to overcome a specific problem, or to fulfill the nutrient needs of a specific crop. These products are considered specialty fertilizers.

Two main approaches to achieving controlled release have been (1) the development of compounds of limited water solubility and (2) the alteration of soluble materials to retard their nutrient release to the soil solution. Nitrogen, the most widely used fertilizer nutrient, is also the most susceptible to loss by either leaching or denitrification. This loss has received the greatest attention.

Controlling the release of nitrogen can be accomplished by (1) adding a physical barrier (coating) to water-soluble materials; (2) using materials of limited water solubility, e.g., metal ammonium phosphates; and (3) using materials of limited water solubility which, during chemical and/or microbiological decomposition, release nutrients in available forms, e.g., methylene ureas. Some common products are listed in Table 5-8.

## Table 5-8
### Analyses of Some Specialty Fertilizer Compounds

| Material | Method of Controlling Release | Nutrient Content |
|---|---|---|
| | | (%) |
| Resin-coated NPK | Coating with resin or plastic | 18-5-11 14-14-14 18-6-12 |
| Sulfur-coated urea | Coating with sulfur | 32-37 N |
| Isobutylidene-diurea | Solubility | 31 N |
| Methylene urea | Solubility | 38-41 N |
| Melamine | Solubility | 66 N |
| Oxamide | Solubility | 28-32 N |
| Triazone | Solubility | 41 N |

## Coated Fertilizers

Urea, because of its high analysis and physical properties,

is the principal nitrogen material used in coated fertilizers. Commercial coating processes use resins, thermoplastics, polymers, and related reactive layer coating (RLC) technology. Sulfur is also used as a coating to slow the release of nitrogen.

### Uncoated Inorganic Materials

These compounds comprise a group of compounds of the general formula $MeNH_4PO_4 \bullet xH_2O$, where "Me" is a divalent metal. All have limited solubility, and several have been developed as multiple nutrient, slow release fertilizers. Magnesium ammonium phosphate is the most common. Water-soluble nitrogen in these materials varies between one and two percent.

### Uncoated Organic Compounds

Two of the most common are the group of methylene urea (MU) products and isobutylidene-diurea (IBDU). The composition and therefore the nitrogen release characteristics of methylene ureas is controlled by the ratio of short to long-chain polymers. In manufacturing, the process is carefully controlled to obtain a suitable balance between the relatively soluble short-chain and more insoluble long-chain polymers.

# NUTRIENT
# CONVERSION FACTORS

In North America, the nutrients in fertilizers are reported in elemental form with the exception of phosphorus and potassium. Figure 5-7 can be used for rapid conversion of phosphorus and potassium to the oxide forms or the reverse.

Several foreign countries report all nutrients in fertilizers on the elemental basis. Most of the scientific literature is also reported this way. Often it is necessary to determine the percentage of an element in a fertilizer material. Table 5-9 provides

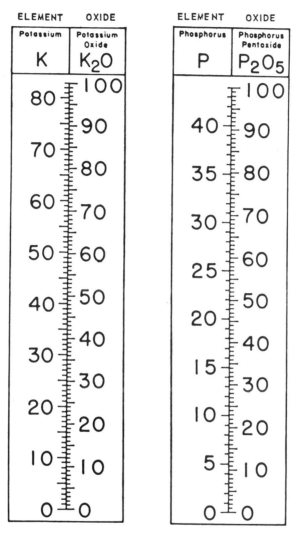

Figure 5-7. Fertilizer conversion chart for changing
oxide to element and vice versa (pounds or percent).

easy conversion factors to make the conversion from the compound to the element and from the element to the compound
for common fertilizer materials.

## Table 5–9
## Conversion Factors

To find the equivalent of one material, A, in terms of another, B, multiply the amount of A by the factor in column "A to B." To find the equivalent of material B in terms of A, multiply the amount of B by the factor in column "B to A."

| A | B | Multiply | |
|---|---|---|---|
| | | A to B | B to A |
| Ammonia ($NH_3$) | Nitrogen (N) | 0.8224 | 1.2159 |
| Nitrate ($NO^3$) | Nitrogen (N) | 0.2259 | 4.4266 |
| Protein (crude) | Nitrogen (N) | 0.1600 | 6.2500 |
| Ammonium nitrate ($NH_4NO^3$) | Nitrogen (N) | 0.3500 | 2.8572 |
| Ammonium sulfate [$(NH_4)_2SO_4$] | Nitrogen (N) | 0.2120 | 4.7168 |
| Calcium nitrate [$Ca(NO_3)_2$] | Nitrogen (N) | 0.1707 | 5.8572 |
| Ammonium calcium nitrate decahydrate [$5Ca(NO_3)_2•NH_4NO_3•10H_2O$] | Nitrogen (N) | 0.1883 | 5.3107 |
| Potassium nitrate ($KNO_3$) | Nitrogen (N) | 0.1386 | 7.2176 |
| Sodium nitrate ($NaNO_3$) | Nitrogen (N) | 0.1648 | 6.0679 |
| Monoammonium phosphate ($NH_4H_2PO_4$) | Nitrogen (N) | 0.1218 | 8.2118 |
| Diammonium phosphate [$(NH_4)_2HPO_4$] | Nitrogen (N) | 0.2121 | 4.7138 |
| Urea | Nitrogen (N) | 0.4665 | 2.1437 |
| Phosphoric acid ($P_2O_5$)[1] | Phosphorus (P) | 0.4364 | 2.2914 |
| Phosphate ($PO_4$) | Phosphorus (P) | 0.3261 | 3.0662 |
| Monoammonium phosphate ($NH_4H_2PO_4$) | Phosphoric acid ($P_2O_5$)[1] | 0.6170 | 1.6207 |
| Diammonium phosphate [$(NH_4)_2HPO_4$] | Phosphoric acid ($P_2O_5$)[1] | 0.5374 | 1.8607 |
| Monocalcium phosphate [$Ca(H_2PO4)_2$] | Phosphoric acid ($P_2O_5$)[1] | 0.6068 | 1.6479 |
| Dicalcium phosphate ($CaHPO_4•2H_2O$) | Phosphoric acid ($P_2O_5$)[1] | 0.4124 | 2.4247 |
| Tricalcium phosphate [$Ca_3(PO_4)_2$] | Phosphoric acid ($P_2O_5$)[1] | 0.4581 | 2.1829 |
| Potash ($K_2O$) | Potassium (K) | 0.8301 | 1.2046 |
| Muriate of potash (KCl) | Potash ($K_2O$) | 0.6317 | 1.5828 |
| Sulfate of potash ($K_2SO_4$) | Potash ($K_2O$) | 0.5405 | 1.8499 |
| Potassium of nitrate ($KNO_3$) | Potash ($K_2O$) | 0.4658 | 2.1466 |
| Potassium carbonate ($K_2CO_3$) | Potash ($K_2O$) | 0.6816 | 1.4672 |
| Gypsum ($CaSO_4•2H_2O$) | Calcium sulfate ($CaSO_4$) | 0.7907 | 1.2647 |
| Gypsum ($CaSO_4•2H_2O$) | Calcium (Ca) | 0.2326 | 4.3000 |
| Gypsum ($CaSO_4•2H_2O$) | Calcium oxide (CaO) | 0.3257 | 3.0702 |

*(Continued)*

## Table 5-9 (Continued)

| A | B | Multiply A to B | B to A |
|---|---|---|---|
| Calcium oxide (CaO) | Calcium (Ca) | 0.7147 | 1.3992 |
| Calcium carbonate (CaCo$_3$) | Calcium (Ca) | 0.4004 | 2.4973 |
| Calcium carbonate (CaCo$_3$) | Calcium oxide (CaO) | 0.5604 | 1.7848 |
| Calcium carbonate (CaCo$_3$) | Calcium hydroxide [Ca(OH)$_2$] | 0.7403 | 1.3508 |
| Calcium hydroxide [Ca(OH)$_2$] | Calcium (Ca) | 0.5409 | 1.8487 |
| Magnesium oxide (MgO) | Magnesium (Mg) | 0.6032 | 1.6579 |
| Magnesium sulfate (MgSO$_4$) | Magnesium (Mg) | 0.2020 | 4.9501 |
| Epsom salts (MgSO$_4 \bullet 7H_2O$) | Magnesium (Mg) | 0.0987 | 10.1356 |
| Sulfate (SO$_4$) | Sulfur (S) | 0.3333 | 3.0000 |
| Ammonium sulfate [(NH$_4$)2SO$_4$] | Sulfur (S) | 0.2426 | 4.1211 |
| Gypsum (CaSO$_4 \bullet 2H_2O$) | Sulfur (S) | 0.1860 | 5.3750 |
| Magnesium sulfate (MgSO$_4$) | Sulfur (S) | 0.3190 | 3.1350 |
| Potassium sulfate (K$_2$SO$_4$) | Sulfur (S) | 0.1837 | 5.4438 |
| Sulfuric acid (H$_2$SO$_4$) | Sulfur (S) | 0.3269 | 3.0587 |
| Borax (Na$_2$B$_4$O$_7 \bullet 10H_2O$) | Boron (B) | 0.1134 | 8.8129 |
| Boron trioxide (B$_2$O$_3$) | Boron (B) | 0.3107 | 3.2181 |
| Sodium tetraborate penta-hydrate (Na$_2$B$_4$O$_7 \bullet 5H_2O$) | Boron (B) | 0.1485 | 6.7315 |
| Sodium tetraborate anhydrous (Na$_2$B$_4$O$_7$) | Boron (B) | 0.2150 | 4.6502 |
| Cobalt nitrate [Co(NO$_3$)$_2 \bullet 6H_2O$] | Cobalt (Co) | 0.2025 | 4.9383 |
| Cobalt sulfate (CoSO$_4 \bullet 7H_2O$) | Cobalt (Co) | 0.2097 | 4.7690 |
| Cobalt sulfate (CoSO$_4$) | Cobalt (Co) | 0.3802 | 2.6299 |
| Copper sulfate (CuSO$_4$) | Copper (Cu) | 0.3981 | 2.5119 |
| Copper sulfate (CuSO$_4 \bullet 5H_2O$) | Copper (Cu) | 0.2545 | 3.9293 |
| Ferric sulfate [Fe$_2$(SO$_4$)$_3$] | Iron (Fe) | 0.2793 | 3.5804 |
| Ferrous sulfate (FeSO$_4$) | Iron (Fe) | 0.3676 | 2.7203 |
| Ferrous sulfate (FeSO$_4 \bullet 7H_2O$) | Iron (Fe) | 0.2009 | 4.9776 |
| Manganese sulfate (MnSO$_4$) | Manganese (Mn) | 0.3638 | 2.7486 |
| Manganese sulfate (MnSO$_4 \bullet 4H_2O$) | Manganese (Mn) | 0.2463 | 4.0602 |
| Sodium molybdate (Na$_2$MoO$_4 \bullet 2H_2O$) | Molybdenum (Mo) | 0.3965 | 2.5218 |
| Sodium nitrate (NaNO$_3$) | Sodium (Na) | 0.2705 | 3.6970 |
| Sodium chloride (NaCl) | Sodium (Na) | 0.3934 | 2.5417 |
| Zinc oxide (ZnO) | Zinc (Zn) | 0.8034 | 1.2447 |
| Zinc sulfate (ZnSO$_4$) | Zinc (Zn) | 0.4050 | 2.4693 |
| Zinc sulfate (ZnSO$_4 \bullet H_2O$) | Zinc (Zn) | 0.3643 | 2.7449 |

[1]Also called phosphoric acid anhydride, phosphorus pentoxide, available phosphoric acid.

# SUPPLEMENTARY READING

1. *Agricultural Anhydrous Ammonia: Technology and Use.* M. H. McVickar, W. P. Martin, I. E. Miles and H. H. Tucker, eds. Soil Sci. Soc. of America. 1966.

2. *Fertilizer Manual.* International Fertilizer Development Center. United Nations Industrial Development Organization. 1979.

3. *Fertilizer Nitrogen, Its Chemistry and Technology,* Second Edition. V. Sauchelli. Reinhold Publishing Corp. 1968.

4. *Fertilizers and Soil Fertility,* Second Edition. U. S. Jones. Reston Publishing Co. 1982.

5. *Fertilizer Technology and Use.* O. P. Engelstad, T. J. Army, J. J. Hanway and V. J. Kilmer, eds. Soils Sci. Soc. of America. 1985.

6. *The Merck Index,* Tenth Edition. M. Windholz, ed. Merck and Co., Inc. 1983.

7. *Micronutrients in Agriculture,* Second Edition. J. J. Mortvedt, F. R. Cox, L. M. Shuman and R. M. Welch, eds. Soil Sci. Soc. of America. 1991.

8. *Nitrogen in Agricultural Soils.* F. J. Stevenson, ed. American Society of Agronomy. 1982.

9. *Potassium in Agriculture.* R. D. Munson, ed. American Society of Agronomy. 1985.

10. *The Role of Phosphorus in Agriculture.* F. E. Khasawneh, E. C. Sample and E. J. Kamprath, eds. American Society of Agronomy. 1980.

11. *Soil Fertility and Fertilizers,* Fourth Edition. S. L. Tisdale, W. L. Nelson, and J. D. Beaton. The Macmillan Company. 1985.

12. *Using Commercial Fertilizer,* Fourth Edition. M. H. McVickar and W. M. Walker. The Interstate Printers & Publishers, Inc. 1978.

# Chapter 6

# *Fertilizer Formulation, Storage, and Handling*

With the rapid technological developments of recent years in the fertilizer trade, the farmer, as well as the fertilizer dealer and distributor, has a wide choice of fertilizer systems available for use. These may be grouped into three broad classifications. The first system has been developed to bulk blend or prescription mix the nutrient requirements for specific soil and crop requirements. This system may use homogeneous products as part of the blend, along with other granular fertilizer materials blended to meet the desired nutrient requirements. Such products should be uniformly sized to minimize segregation in storage and handling. Second is the system that makes direct application of homogeneous products. These products are available in a number of grades and ratios that meet many soil requirements. Storage and handling are usually in bulk form, but bagged products may also be obtained. The third is the fluid fertilizer system that has products ranging from clear liquid solutions to suspensions. Its assets are ease of handling, uniform composition, and adaptability to additions of herbicides and insecticides.

The adoption of these systems has been hastened because of the development of new products and because they provide costs savings at all levels from manufacture to transportation, storage, and application. With new technological breakthroughs, substitutions or shifts from one system to another will occur.

This is well illustrated by the development of superphosphoric acid, which has stimulated the expansion of the fluid fertilizer business. Suspension fertilizers can now be formulated that have a higher nutrient content than that of clear solutions and can include a uniform mix of plant nutrients.

Fertilizer suppliers have several options as to the type of production unit they operate and the materials they have available. Factors influencing their choices include capital assets, size of distribution area, availability of raw materials, agronomic suitability of certain products, and personal preferences.

# FORMULATION

## BULK BLENDS

Bulk blends are physical mixes of two or more dry fertilizer materials. The bulk blend plant receives fertilizer products from a basic producer, stores them and blends them together as needed in some type of mixing device. Some of the materials more commonly used to make blends are ammonium nitrate, ammonium sulfate, diammonium phosphate, mono ammonium phosphate, urea, and potassium materials. The blends may be taken directly to the field and spread, or, in some cases, they may be bagged. One of the potential problems with some blends is segregation or separation of one component or raw material from another. The principal factor contributing to segregation is the use of materials of different sizes. Most basic suppliers of fertilizer materials produce standard size grades, to help eliminate problems in blending.

Blenders must exercise care in the selection of raw materials. Also, blenders should avoid allowing piles of finished product to cone, either in storage or when loaded into a truck or trailer. In coning, the larger materials run to the outside of a pile, while the smaller materials stay in the center. Coning can be largely

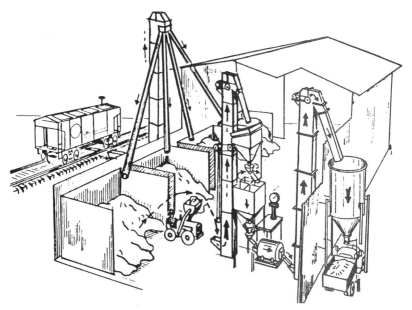

Figure 6-1. Schematic drawing of a bulk blending plant.

overcome with the use of a flexible spout. Particle shape and density contribute little to segregation problems.

Micronutrients can be added to blends, but a granular form having the same size range as other components is recommended. Another method of incorporating micronutrients is by spraying a solution of the elements on the blend during mixing.

## HOMOGENOUS DRY PRODUCTS

The unique characteristic of a homogeneous product is that each granule or pellet has the same analysis. Homogeneous products are usually manufactured in large fertilizer plants and are supplied to the resellers where they may be used in blends or applied directly to the soil. These facilities utilize ammonia, sulfuric and phosphoric acids, and other raw materials.

Some common grades are 18-46-0, 16-20-0, 11-52-0, 6-20-20, and 12-12-12. In addition to the three primary nutrients, nitrogen, phosphorus, and potassium, it is also possible to include micronutrients as a guaranteed constituent in the product.

One of the problems with homogeneous products is that specific grades are not produced to meet every soil or crop condition. Not enough or excessive amounts of one nutrient may be applied in order to meet the requirement of another nutrient.

## FLUIDS

Fluid fertilizers include both clear liquids and suspensions. Liquids include nitrogen solutions, phosphoric acid, and liquid mixes. Nitrogen solutions have already been discussed in Chapter 5.

Liquid mixes are produced by neutralizing phosphoric acid with ammonia. If orthophosphoric acid is used, the usual product will be an 8-24-0 grade (Table 6-1). Ammoniation can be accomplished in mild steel tanks. Where ammonia and the acid come in contact and are mixed, stainless steel is preferred. Considerable heat is released during ammoniation, and the solution will be hot. If potassium is desired, it should be added

|  | **Table** |
| :--- | ---: |
|  | **Properties of Ammonium** |
| Grade | 8-24-0 |
| Acid used in production | Orthophosphoric |
| Percent N by weight | 8 |
| Percent $P_2O_5$ by weight | 24 |
| Density, lbs./gal. @ 60°F | 10.5 |
| N content, lbs./gal. | 0.84 |
| $P_2O_5$ content, lbs./gal. | 2.52 |
| Polyphosphates, % of total $P_2O_5$ | none |
| Viscosity, CP @ 75°F | — |
| Safe storage temperature, °F | 12 |
| pH | 6.4–6.6 |

at this time, as the introduction of potassium will lower the solution temperature.

If superphosphoric acid (68 to 76 percent $P_2O_5$, 50 to 75 percent of which is present as polyphosphate) is ammoniated, it is possible to produce a stable 10-34-0 solution. The higher analysis of this solution is a result of the greater solubility of the pyro-, tri-poly, and more condensed phosphates. As condensed species are rather unstable at high temperatures, it is necessary to cool 10-34-0 during production. At high temperatures, hydrolysis of the condensed phosphate occurs with reversion to the ortho form as illustrated in Figure 6-2. When this occurs, the "salting out" temperature is raised, making the solution unstable. Therefore, cooling the solution to below 100°F within an hour after production is recommended.

Low poly content (20 percent) superphosphoric acid contains significant quantities of metal ions such as magnesium, iron, and aluminum. When the acid is ammoniated in the pipe reactor process, developed at TVA, the heat of reaction of the acid and the ammonia drives off additional combined water and forms a large proportion of condensed phosphate (70 to 80

**6-1**
**Phosphate Clear Liquids**

| 9-30-0 | 10-34-0 | 11-37-0 |
|---|---|---|
| Superphosphoric | Superphosphoric | Superphosphoric |
| 9 | 10 | 11 |
| 30 | 34 | 37 |
| 11.3 | 11.4 | 11.7 |
| 1.02 | 1.14 | 1.29 |
| 3.39 | 3.87 | 4.33 |
| 40-45 | >50 | >65 |
| – | 73 | 80 |
| 0 | below 0 | 0 |
| 6.2-6.6 | 5.8-6.1 | 5.8-6.2 |

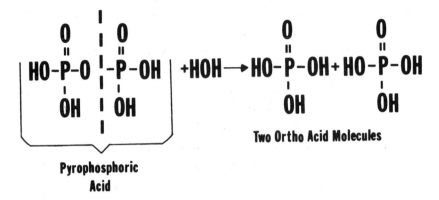

Figure 6-2. Hydrolysis of pyrophosphoric acid.

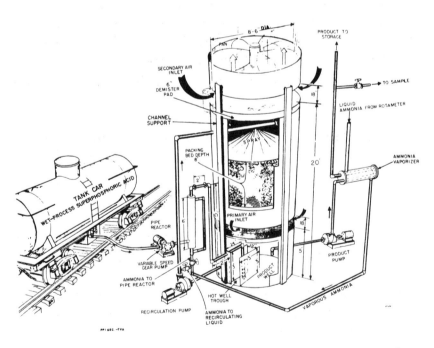

Figure 6-3. Reactor for the production of high polyphosphate 10-34-0 from low polyphosphoric acid.

percent) in the final 10-34-0 product. A sketch of a pipe reactor unit is shown in Figure 6-3.

Metal salts of long-chain condensed phosphates are more soluble than ortho- and pyrophosphates. Consequently, the metals are in solution and do not cause precipitation or sludge problems in the 10-34-0 grade.

Although rather high analysis N-P grades can be produced as clear liquids, the addition of other fertilizer materials raises the salting out temperature, thus limiting the nutrient content and multi-nutrient grades. There has been a constant increase in the grades of fluid fertilizers with the progression from orthophosphate to polyphosphate acids.

Due to limitations in producing high analysis potassium, sulfur, and micronutrient solutions, some dealers have turned to suspensions.

Suspensions are saturated solutions containing suspended crystals of plant nutrients or other materials. Usually, an agent such as an attapulgite-type gelling clay is used to suspend the undissolved crystals throughout the liquid medium.

Some of the advantages suspensions have over liquid mix fertilizers are:

1. Higher analysis suspension grades can have as much as twice the nutrient content of clear liquid mixtures.

2. Larger amounts of secondary and micronutrients and other elements can be used in suspensions.

Typical grades of clear liquid mixes and suspension fertilizers are shown in Table 6-2. Suspensions can be produced through four principal reactions: combining orthophosphoric acid with ammonia, polyphosphoric acid with ammonia, solid ammonium phosphates with ammonia, and diammonium phosphate with sulfuric acid.

A relatively recent development in suspension production has been the use of dry ammonium phosphate, particularly

Figure 6-4. Ammonium polyphosphate production facility reacting anhydrous ammonia with super phosphoric acid to produce hydrolysis clear 10-34-0 solutions. The plume or steam is produced by the heat of reaction between acid and base.

Table 6–2
Typical Grades of Some Clear Liquid Mixes and
Suspension Fertilizers for Different Ratios

| Ratio | Grade | |
|---|---|---|
| | Clear Liquid Mix | Suspension |
| 3:1:0 | 24-8-0 | 27-9-0 |
| 2:1:0 | 22-11-0 | 26-13-0 |
| 1:1:0 | 19-19-0 | 21-21-0 |
| 1:1:1 | 8-8-8 | 15-15-15 |
| 1:2:2 | 5-10-10 | 10-20-20 |
| 1:3:1 | 7-21-7 | 10-30-10 |
| 1:3:2 | 5-15-10 | 9-27-18 |
| 1:3:3 | 3-9-9 | 7-21-21 |

monoammonium phosphate. This system may have certain advantages, as it permits the producer to use inexpensive dry material storage. Also, the phosphate in dry ammonium phosphate may be less costly than that in phosphoric acid.

The dry ammonium phosphate also permits a dealer to market both dry and liquid materials with only a modest increase in capital cost.

To produce suspensions from monoammonium phosphate, one must have a mixing tank, an agitator, and a means of sparging in ammonia and introducing clay. It is important to understand the relationship between ammonium nitrogen and phosphate to produce good quality suspensions.

There may be application problems with suspensions. Due to the presence of the finely divided crystals in the fluid, buildup and clogging can occur. Constant agitation is needed to keep the materials in suspension.

# STORAGE AND HANDLING

To bring efficiency to fertilizer production and distribution, proper and safe storage must be an integral part of the system. Fertilizer plants should operate throughout the year to be efficient. Storage by the producer is necessary to keep the facilities operating, and storage by the dealer and farmer is necessary to ensure having a supply available when needed. Those who contemplate storage should make a thorough study of their needs and the facilities available.

## DRY MATERIALS

*Ammonium nitrate* — It is an excellent fertilizer material that presents no hazard when good storage and proper handling procedures are observed. Precautions to be taken are:

1. Keep it away from open flame.

2. Avoid contaminating it with foreign matter.

3. In case of fire, flood the area with water.

4. Burn empty bags out of doors.

5. Sweep up and dispose of all contaminated material.

6. Do not store in close proximity to steam pipes or radiators.

7. Keep it separate from other materials stored in the same warehouse, especially combustible materials and urea.

Storage of dry urea is incompatible with dry ammonium nitrate. In the presence of urea, atmospheric moisture will be taken up at lower relative humidity (18 percent). Critical relative humidity for individual or combination products is presented in Table 6-3.

Ammonium nitrate has a critical relative humidity of 59.4 percent at 86°F, and under humid conditions it tends to absorb moisture. In areas of high humidity where ammonium nitrate is manufactured, dehumidified storage may be necessary. At small retail outlets and bulk blend plants, a tight bin and a polyethylene cover sheet can be used for storage. Because this product is corrosive, a concrete or wooden storage structure is preferred.

Whenever ammonium nitrate accidentally becomes contaminated with combustible materials, it should be disposed of in a safe manner.

*Urea* — Storage of urea is incompatible with ammonium nitrate. The physical handling and storage of urea is very similar to ammonium nitrate except it is less corrosive. The critical relative humidity is 72 percent at 86°F. Thus, it is hygroscopic.

*Ammonium sulfate* — This material is safe and easy to store. Because of its high critical relative humidity of 81 percent at 86°F, storage problems are infrequent.

The product is corrosive, so concrete or wooden storage structures are preferred.

## Table 6–3
## The Hygroscopicity of Some Pure Compounds
## That May Occur in Fertilizers and Their Mixtures.

| Fertilizer Salt | Critical Relative Humidity at Which the Salt or its Mixture with Other Salts Begins to Absorb Moisture at 86°F. [a] | | | | | | | | | | | | | |
|---|---|---|---|---|---|---|---|---|---|---|---|---|---|---|
| Calcium nitrate $Ca(NO_3)_2 \cdot 4H_2O$ | | | | | | | | | | | | | | 47 |
| Ammonium nitrate $NH_4NO_3$ | | | | | | | | | | | | | 59 | 24[b] |
| Sodium nitrate $NaNO_3$ | | | | | | | | | | | | 72 | 46[b] | 38[b] |
| Urea $CO(NH_2)_2$ | | | | | | | | | | | 73 | 46[b] | 18[b] | 67[c] |
| Sodium chloride $NaCl$ | | | | | | | | | | 75 | 53[b] | 68[b] | d | |
| Ammonium chloride $NH_4Cl$ | | | | | | | | | 77 | 69[b] | 58[b] | 42-52[e] | 51[b] | |
| Ammonium sulfate $(NH_4)_2SO_4$ | | | | | | | | 79 | 71[b] | 56[b] | d | 62[b] | | |
| Potassium chloride $KCl$ | | | | | | | 84 | d | 74[b] | 72[b] | 60[b] | d | d | e |
| Potassium nitrate $KNO_3$ | | | | | | 91 | 79[b] | 62-69[e,f] | 55-68[e,f] | 61-67[e] | 65[b] | 65[b] | 60[b,f] | 31[b] |
| Ammonium phosphate $NH_4H_2PO_4$ | | | | | 92 | 60[e,f] | d | 76[b] | 74[b] | 65[b] | 49-64[e] | 58[b] | d | |
| Potassium phosphate $KH_2PO_4$ | | | | 93 | 91[b] | 90[b] | 83[b] | d | 73[e,f] | 70[b] | 63[e] | d | d | |
| Calcium phosphate $Ca(H_2PO_4)_2$ | | | 94 | 91[b] | 89[b] | 44-88[e] | 78[e] | d | 74[e] | 65[b] | 36-68[e] | 26-53[e] | 46[b] | |
| Potassium sulfate $K_2SO_4$ | | 96 | d | 94[b] | 79[e,f] | 88[b] | 83[b] | 81[b] | 71[e,f] | 72[b] | d | d | d | |
| Calcium sulfate $CaSO_4$ | 93-94[e] | 77-92[e] | 30-92[e] | | | | | | | | | | | |

a   The first value is the approximate critical relative humidity of the salt alone. Read down or across to find the critical relative humidity of its mixture with most of the other salts listed.
b   Nonreacting salt pairs. Data for solution saturated with both salts.
c   The double salt, $Ca(NO_3)_2 \cdot 4CO(NH_2)_2$.
d   Unstable salt pair of reciprocal salt pairs. See stable salt pair.
e   Stable salt pair of reciprocal salt pairs.
f   Solid phases include solid solutions.

Source: *1992 Farm Chemicals Handbook*, p. B18.

*Phosphorus and potassium materials* — Except under extremely adverse conditions, the ammonium phosphates, super phosphates, and potassium materials require no specialized storage. Like most fertilizers, they tend to be corrosive, so concrete and wood are preferred materials for storage structures. Due to the density, large piles or bags or these materials stacked excessively high may cause "setting up," but the lumps are easily broken.

## FLUID MATERIALS

*Anhydrous Ammonia* — This material is widely used in manufacturing and for direct application. It is potentially hazardous,
but handling procedures and safety precautions are well known
and must be observed. Anhydrous ammonia storage and handling will not be discussed here, but it is recommended that
anyone contemplating handling or using this product become
acquainted with its characteristics and proper handling procedures. A good source of information is The Fertilizer Institute's
manual *Agricultural Ammonia Safety.*

*Aqua Ammonia* — A 20 percent solution of ammonia-nitrogen
in water that has a very low gauge pressure. Ammonia vapor is
constantly leaving the solution. Therefore, a pressure-vacuum
relief valve must be installed on storage tanks. This is also true
of other solutions which contain free ammonia.

*Urea-Ammonium Nitrate Solution* — Commonly used non-
pressure solutions are made from urea, ammonium nitrate, and
water. Standard grades are 28 percent and 32 percent nitrogen.
The former is used during cold weather since it has a much
lower salting out temperature. These solutions are used for
direct application and for making multi-nutrient liquid fertilizers
by combining them with other fertilizer materials.

*Ammonium Nitrate Solution* — The common non-pressure solution of ammonium nitrate in water is usually standardized at
20 percent nitrogen content. It is used for direct application or
for making multi-nutrient liquid fertilizers. Solutions containing
higher concentrations of ammonium nitrate are used only for
manufacturing purposes, since they need to be kept hot to
prevent salting out.

*Urea Solution* — The common urea solution in water contains
23 percent nitrogen. It is used for direct application or for
making multi-nutrient liquid fertilizers. Some solutions containing higher concentrations are used for manufacturing purposes

where, as with ammonium nitrate solutions, they must be kept hot to prevent salting out.

*Phosphoric and Superphosphoric Acids* – These acids are widely used by both fertilizer manufacturers and formulators. Both are corrosive, although superphosphoric acid is somewhat less corrosive than orthophosphoric acid. Rubber-lined, stainless steel or plastic tanks and plumbing are necessary for ortho-phosphoric acid and are recommended for superphosphoric acid.

Superphosphoric acid is hygroscopic and will form a thin layer of more corrosive orthophosphoric acid on its surface, if allowed to absorb moisture. This should be prevented by having a silica-gel breather installed to prevent moisture from entering the tank.

The viscosity of superphosphoric acid increases as the temperature decreases.

Although centrifugal pumps can move hot superphosphoric acid, most operators prefer positive displacement, gear, screw

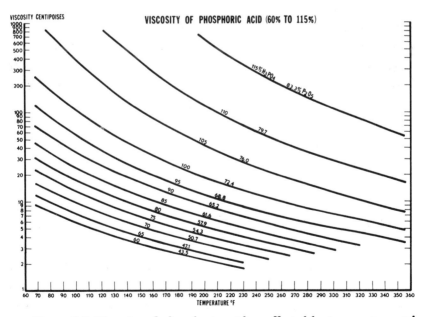

Figure 6-5. Viscosity of phosphoric acid as affected by temperature and concentration.

or positive vane pumps. Wetted parts should be made of corrosion-resistant alloy. Mechanical seals can be used if the acid
is clean. Long lines should be provided with external jacketing
or steam tracing, to prevent blockages due to gelling or freezing.

Phosphoric or other acids should not come in contact with
skin or eyes. These acids are not extremely hazardous, and
prompt washing with copious quantities of water is an effective
remedy. Superphosphoric acid with 76 to 83 percent $P_2O_5$ is a
strong dehydrating agent; therefore, it has a greater tendency
to cause blistering than less concentrated grades.

*Clear Liquid and Fluid Suspensions* — Clear liquid fertilizers
are easy to handle. If they are neutral solutions, mild steel
storage can be utilized and most conventional pumping systems
can be used.

Suspensions store best in vertical mild steel or plastic tanks
equipped for air sparging. The life of a steel tank can be extended
by coating the inside with a material such as an epoxy base
paint. The tanks may have flat or cone bottoms, although it is
easier to resuspend materials that settle in cone-bottom tanks.
When suspensions are not of good quality, solids may tend to
collect on the bottom outside walls of a flat-bottom tank where
they cannot be resuspended by air sparging.

Air sparging is accomplished when an air sparging ring is
mounted on the bottom of the storage tank in such a manner
that agitation will occur throughout the tank when the sparging
lines are pressurized. The holes in the sparging ring point
downward. A compressor with a 300-gallon tank pressurized to
100 to 125 pounds per square inch should be adequate for
most storage. Sparging once or twice a week should be adequate
for most suspensions. Positive displacement pumps are preferred
for suspensions.

*Sulfuric Acid* — This acid is used extensively in the fertilizer
industry. It is mostly used by basic producers in the manufacture
of phosphate compounds and ammonium sulfate. Occasionally,

Figure 6-6. Modern liquid fertilizer facility designed to meet environmental compliance and production efficiency standards: A four inch closed air loop system pneumatically transfers dry fertilizer to overhead bins. A venturi water scrubbing feature for dust suppression is included. The sloping concrete pad in the loadout area contains spills. Surrounding the tank farm at left are secondary containment walls to prevent fertilizer escape into the environment in the event of a ruptured storage tank.

it may be used by a dealer or farmer to correct a specific soil problem.

Sulfuric acid has inherent hazards, yet it can be handled and used safely by following appropriate precautionary methods. Diligent use of safety apparel, protective equipment, and properly engineered handling and storage equipment makes it possible to operate without accidents.

Storage areas must have facilities for containment and recovery of spills. Storage tanks should be equipped with vents to maintain the tank at atmospheric pressure. Metal catwalks should be provided for working on top of tanks.

Limestone or other alkaline products can be used to neutralize any unrecovered spilled acid.

Although sulfuric acid is not flammable, it should not be stored near organic materials, nitrates, carbides, chlorates or metal powders. Contact between concentrated sulfuric acid and these materials may cause ignition. As the acid is diluted with water it becomes more corrosive, and the use of a nonreactive liner for the storage tank is advisable.

*Sulfur Materials for Formulation of Liquids* — These materials include ammonium bisulfite, ammonium thiosulfate, potassium thiosulfate, and ammonium polysulfide. For more details on these materials, see Chapter 5.

A major problem of liquid fertilizers is corrosion, and this demands familiarity with the characteristics of various solutions and the composition of storage containers. Storage vessels, pipelines, valves, and fittings of all kinds should not contain materials that are degraded when they come in contact with the solution.

## SUPPLEMENTARY READING

1. *Agricultural Anhydrous Ammonia: Technology and Use.* M. H. McVickar, W. P. Martin, I. E. Miles and H. H. Tucker, eds. Soil Sci. Soc. of America. 1966.

2. *Code of Federal Registrations,* CFR 29 Part 1910.109-Occupational Safety and Health Administration. Washington D.C. 1992.

3. *Fertilizer Nitrogen, Its Chemistry and Technology,* Second Edition. V. Sauchelli. Reinhold Publishing Corp. 1968.

4. *Fertilizers and Soil Fertility,* Second Edition. U. S. Jones. Reston Publishing Co. 1982.

5. *Fertilizer Technology and Use,* Third Edition. O. P. Engelstad, eds. Soil Sci. Soc. of America. 1985.

6. *Fluid Fertilizer Manual,* Agricultural Retailers Association. 1987.

7. *Nitrogen in Agricultural Soils.* F. J. Stevenson, ed. American Society of Agronomy. 1982.

8. *Physical Properties of Fertilizers and Methods for Measuring Them.* G. Hoffmeister. National Fertilizer Development Center. Tennessee Valley Authority. 1979.

9. *Potassium in Agriculture.* R. D. Munson, ed. American Society of Agronomy. 1985.

10. *Prevention of Sulfur Fires and Explosions.* American National Standard-NSPA. Quincy, MA. 1988.

11. *The Role of Phosphorus in Agriculture.* F. E. Khasawneh, E. C. Sample and E. J. Kamprath, eds. American Society of Agronomy. 1980.

12. *Safety Requirements for the Storage and Handling of Anhydrous Ammonia.* American National Standards Institute. ANSI K 61.1. 1989.

13. *Soil Fertility and Fertilizers,* Fourth Edition. S. L. Tisdale, W. L. Nelson and J. D. Beaton. The Macmillian Company. 1985.

14. *Storage of Ammonium Nitrate.* NFPA No. 490. National Fire Protection Association. 1974.

# Chapter 7

# *Methods of Applying Fertilizer*

Fertilizers are used to provide nutrients that are not present in soil in amounts necessary to meet the needs of the growing crop. When choosing methods of application, growers should consider the following:

1. Rooting characteristic of the crop to be planted.

2. Crop demand for various nutrients at different stages of growth.

3. Physical and chemical characteristics of the soil.

4. Physical and chemical characteristics of the fertilizer material to be applied.

5. Availability of moisture.

6. Type of irrigation system used if irrigation is the only, or major, source of water.

7. Frequency and rate of irrigation water to be applied.

Crop production in the western states often requires multiple applications of fertilizers during the season. Several methods of applying fertilizer may be employed in the same field. For example, lettuce fields may receive a preplant broadcast application of nitrogen and phosphorus. A sufficient quantity of phosphorus, a non-mobile nutrient, can be applied preplant to satisfy the crop's needs for the entire season. Considering the mobility of nitrate, lettuce's shallow root system and low early season demand for nitrogen, and the frequency of irrigation used in lettuce, the majority of the total nitrogen requirement should not be applied preplant. As the crop grows, additional nitrogen is injected into the beds near the row, added to the irrigation water, or perhaps both. By contrast, a dryland grain crop may receive only a preplant nitrogen application.

Fertilizers added to the soil undergo transformations that may change their chemical or positional availability. The methods of application are directly related to crop nutrient utilization and the changes the nutrients undergo in the soil. The appropriate method of application should be based on consideration of the factors listed above and be as economic, accurate, and efficient as possible.

This chapter discusses different methods of fertilizer application, the equipment used, and the benefits and limitations of each application technique.

# PREPLANT APPLICATIONS

## BROADCAST

The broadcast method of applying fertilizers consists of uniformly distributing dry or liquid materials over the soil surface. Broadcast applied fertilizer may be either mechanically worked into the soil or left on the soil surface to be incorporated by rainfall or irrigation.

*Drop spreader* – The simplest applicator for dry fertilizers is an inverted triangle-shaped hopper mounted between two wheels. Fertilizer is distributed through adjustable openings in the bottom of the hopper. Its small size limits the fertilizer load and restricts its use to small fields or to areas where large application equipment cannot enter, such as orchards and vineyards.

Patterned after the simple fertilizer spreader is a folding unit that holds a large amount of fertilizer and spreads a wider swath. The unit folds forward for easy transport to the field. When properly calibrated, very accurate application rates can be attained. Both the simple fertilizer spreader and the folding unit are pulled by truck or tractor.

*Spinning disc spreader* – Small bulk spreaders are available which consist of a bin mounted on a two- or four-wheel trailer frame and pulled by a tractor or truck (Figure 7-1). These spreaders can be pulled across fields at speeds up to 20 mph. The fertilizer is usually spread by a horizontal spinning disc in a 20- to 40-foot swath. The operator must exercise care to prevent skips or overlaps of fertilizer.

*Self-propelled spreader* – This bulk fertilizer spreader consists of a large bin mounted on a large truck or a special three- or four-wheel vehicle equipped with flotation tires to reduce soil

Figure 7-1. A tandem bulk fertilizer spreader that holds up to 10 tons of dry material.

compaction (Figure 7-2). The bin can hold from 7 to 10 tons of fertilizer material.

Self-propelled spreaders vary in sophistication from simple horizontal spinning disc applicators to elaborate air flow applicators which are capable of applying more than one fertilizer material at the same time (Figure 7-3). Some models are equipped with computerized calibration features which automatically adjust rates based on ground speed. Bulk spreading equipment can travel across fields at speeds up to 35 mph and can spread fertilizers in a swath up to 60 feet wide. Some rigs are equipped with foam marker devices which aid in precision application. Proper calibration of the spreader and ground speed, and care on the part of the operator, permits accurate, fast and economical application of fertilizer.

*Liquid spreaders* – The basic requirements for a liquid fertilizer broadcast applicator consist of a tank, pressure gauge and regulator, pump, pipes, hoses, fittings, nozzles, and a boom. The applicator can be mounted on a truck, flotation vehicle or

**Figure 7-2. A three-wheeled, 10-ton bulk fertilizer spreader equipped with flotation tires.**

**Figure 7-3.** Modern air flow spreader capable of applying three different materials at one time. Computerized calibration automatically compensates for variation in ground speed. Foam marker capabilities minimize overlaps or skips.

trailer, or directly on a tractor. The speed of application is determined by the rate of flow.

*Manure spreaders* – Broadcast spreading is the most common method of applying manure. Livestock manure is generally applied at rates of 10-20 tons per acre. Lower rates are usually used for poultry manure. Application equipment usually consists simply of a large bin equipped with a belt conveyor or auger that directs solid material to large horizontal spinning discs. Application problems commonly associated with manure are poor distribution uniformity and loss of nitrogen from surface volatility. Both problems can be reduced by soil incorporation.

## INJECTION

Injection refers to placing fertilizers below the soil surface, which is accomplished by using tool bar-mounted knives or shank openers. Drop pipes for liquids, or flexible tubes for dries, deliver materials into the channels made by the opening tools. All fertilizers that can be broadcast on the soil surface can also be injected. Fertilizers are usually injected after plowing, discing, or when furrowing out when the soil is loose and crop residues are well dispersed.

Injection applications offer several potential advantages over surface broadcast applications. Certain nitrogen fertilizers may be subject to gaseous losses when left on the soil surface. Soil injection eliminates these losses. If soils are subject to wind or rain erosion, injection helps to prevent nutrient losses by placing fertilizers below the zone subject to erosion. Immobile nutrients, such as phosphorus, potassium, and some forms of micronutrients, can be physically placed directly into the crop root zone. For these fertilizer materials injection may be more efficient than broadcast for two reasons. First, less fertilizer is in contact with the soil resulting in less fixation and enhanced nutrient availability. Second, fertilizers may be placed deep enough where soil moisture is constant and feeder root activity is greatest. Soil injection is particularly well suited to row or vegetable crops where a limited amount of the soil is explored by crop roots. Placement directly into the root zone reduces fertilizer availability to weeds and minimizes application in "unused" portions of the soil.

Injection application also presents several disadvantages. Power requirements are greater than for surface broadcast application because of the implements being pulled through the soil. Fewer acres can be treated in a given period of time with injection applications. If applied after bedding-up on very wet or heavy soils, injection tools may disrupt the physical integrity of the beds. Mats of straw or other plant matter may accumulate in front of the injection knives when significant crop residues are present.

Injection of liquid fertilizers requires a tank, a pump, a pressure gauge and regulator, hoses, fittings, tool bars and injection knives or shanks. It should be noted that special high-pressure equipment is required for injecting anhydrous ammonia. Because the power requirement for injecting fertilizers is great, a tractor must be used. Tanks for liquid fertilizers, particularly anhydrous ammonia, may be mounted on a tool bar or saddle-mounted, front-end-mounted or mounted on a trailer and towed behind the tractor.

Equipment for injecting dry fertilizers is less elaborate. Usually a large hopper is mounted on the rear tool bar. The metering wheels on the hopper are driven by a chain drive from the tractor wheel, a ground drive unit or a hydraulic motor. Drop tubes connecting the hopper to the injection shanks allow the placement of the fertilizer in the soil.

Anhydrous ammonia and aqua ammonia are usually applied in bands 12 inches or more apart. Modern high-powered tractors permit use of tool bars 24 feet wide, or greater, for the application of these fertilizers. Liquid and dry fertilizers are usually applied preplant at the time furrows and beds are formed. The injection shanks are placed above the ground level about halfway between the point of the lister shovel and the top of the bed. The fertilizer is metered into the soil as it is thrown into the bed. Bed shapers, either included as part of the total injection-bedding operation or pulled through the field after injecting is finished, leave the field ready for planting.

# APPLICATIONS AT PLANTING

Fertilizers applied at planting may be placed with the seed, below and/or to the side of the seed, or on the soil surface above the seed line. Consideration of crop, soil and environmental conditions, method of irrigation, and equipment constraints dictate the preferred method.

## SUBSURFACE BANDING

Placement of the fertilizer directly with the seed is commonly referred to as "pop-up" application. Band placement near the seed has a similar purpose and will be discussed later. Fertilizer placed in either manner is called "starter." The purpose of starter fertilizers is to stimulate rapid emergence and stand establishment by providing a fertilizer source that is immediately available to the germinating plants. A 1-3-1 or 1-4-1 ratio of N-$P_2O_5$-$K_2O$ is generally recommended.

Grain crops have been observed to respond favorably to fertilizers placed with the seed. The fertilizer hopper is attached to the grain drill, and the fertilizer is metered out according to predetermined rates. Newer grain drills have fertilizer hoppers that can hold up to 1,800 pounds of dry fertilizer. Low salt-index liquid fertilizers, such as 10-34-0, have been used safely as starter fertilizers on field, row, and vegetable crops. Equipment requirements for liquid starter applications are presented in the next section. Liquid fertilizer may also be added to transplant water.

Growers should only use low rates of starter fertilizers, particularly for pop-up applications. Fertilizer rates may be increased as distance from the seed increases. Table 7-1 shows the amount of N plus $K_2O$ that agronomists generally agree is safe to use.

Fertilizer sources which should not be placed with the seed include urea, diammonium phosphate, aqua ammonia, ammo-

**Table 7-1**
**Maximum Cumulative Amounts of N Plus K**
**Based on Distance From Seed**

| Placement | Sandy Soils | Fine Textured Soils |
|---|---|---|
|  | *(lbs./A)* | *(lbs./A)* |
| In contact with seed | 5 | 5-8 |
| $\frac{1}{4}$ to $\frac{1}{2}$ inch from seed | 8 | 7-15 |
| 1 to 2 inches from seed | 15 | 20-40 |
| Over 2 inches from seed | 20 plus | 40 plus |

Rates based on 38-40" rows. Rates should be adjusted based on row spacing and local guidelines.

nium polysulfide, ammonium thiosulfate, and any others which may evolve free ammonia under certain conditions. Free ammonia can be very toxic to seedlings. Fertilizer placement with the seed is not recommended when the ECe approaches the salinity threshold for the crop to be grown (see Tables 2-7 through 2-10).

The most common method of banding at-planting is placement of the fertilizer material below and to the side of the seed. Advantages include non-disruption of the seed line, less concern for salt or ammonia toxicity, potential to use higher rates and deeper placement to provide all or most of the crop's nutrient requirement.

Applicators for banding dry fertilizers on row crops are usually mounted on the same tool bar or bed sled as the seeder and are powered by the same drive that powers the seeder. A tube connected to the fertilizer hopper delivers the fertilizer to a furrow opened by a shoe or disc. The tank for liquid fertilizers may be on the tool bar, or saddle-mounted on the tractor. The injection shank of liquid applicators is usually either attached to the seeder or set to inject the fertilizer ahead of the seeder. The band applicator can be set to place the fertilizer in bands at any depth or position relative to the seed.

## SURFACE BANDING

In some situations, liquid fertilizers may be applied to the soil surface in a narrow band directly over the seed line. This may be accomplished by low pressure dribble applications, or higher pressure directed sprays with the spray nozzle being positioned very close to the soil surface. Surface band applications should usually be reserved to where sprinkler irrigation or rainfall is adequate to move the fertilizer into the root zone and to dilute fertilizer salts. Acidic fertilizers as bed-top sprays are sometimes used to provide the dual function of starter fertilizer and anticrustant. The acid serves to dissolve calcium carbonate crusts that impede crop emergence. To be most effective, surface bands should be applied at planting to ensure placement directly over the seed line.

# POST-EMERGENCE APPLICATIONS

## SIDE DRESSING

This method of fertilization refers to the placement of fertilizers beside the crop rows. These applications may be made at the same time the rows are cultivated. Depending on the soil type, the effects of irrigation and the crop, one or several side dress applications may be desirable. Side dress applications are an efficient means of providing the bulk of a crop's nitrogen requirements. By timing applications close to periods of peak nutrient demand, early season leaching losses may be minimized. Side dressing may also be effective for deep placement of phosphorus and/or potassium close to when it is needed, thus lessening the time for soil fixation to occur.

Both liquid and dry fertilizers may be side dressed. The kind of fertilizer being side dressed determines its placement relative to the crop row. Side-banded phosphorus and potassium fertilizers must be placed close enough to the row and sufficiently deep to be available to the crop's roots. The placement of most nitrogen fertilizers is not as critical because nitrate will move upward and away from the root zone as irrigation water moves from the furrow row to the bed surface. Anhydrous ammonia or aqua ammonia should be injected farther out and sufficiently deep to prevent gaseous losses and/or crop injury. Plants are sensitive to ammonia. Increasing the distance between the crop row and these materials allows for diffusion into the soil, thereby reducing the ammonia concentration in the root zone.

Equipment for side dressing dry fertilizer usually consists

of two large hoppers mounted on tool bars on either side of a tractor, ahead of the operator, or one or more hoppers mounted on a three-point hitch tool bar behind the operator. The metering device is tractor-, ground- or power-driven. Flexible tubing connects the hopper to spouts mounted ahead of, between, or behind discs or shovels. The desired amount of fertilizer is metered into a small furrow and is covered by the cultivating unit. Liquid side-dressed fertilizer tanks are usually saddle-mounted and are connected to a pump and flow control. Hoses from the control are connected to the injection shanks or spouts mounted ahead of, between, or behind discs or shovels. Fertilizer flows into the shank channels or small furrows and is covered by the cultivating unit.

## TOP DRESSING

This method of application consists of spreading fertilizers on the soil surface after crop emergence. Field crops such as small grains, pastures, rangeland, and alfalfa can be top dressed once or several times during the growing season. The same equipment used for preplant broadcast applications of dry and liquid fertilizer may be used for top dressing. Vehicles used for top-dressed applications should be equipped with flotation tires to minimize crop damage and soil compaction.

Aerial applications may be used to eliminate crop damage from equipment and can effectively cover a large number of acres in a short amount of time. Aerial top-dressed applications are commonly made on rice because ground equipment cannot enter the flooded fields.

## WATER-RUN APPLICATIONS

Savings in time, labor, equipment, and fuel costs are advantages of water-run fertilizer applications. Applications may be preplant or post-emergence, using either liquid or dry fertilizer materials. The plant nutrients should not be introduced into the system at the initiation of the irrigation set, as non-uniform distribution and excessive leaching of certain nutrients may occur. For water-run applications, fertilizer distribution uniformity is determined by irrigation distribution uniformity. Best results may be obtained when the fertilizer application is timed to enter the system toward the middle of the set and to terminate shortly before the set is completed. Application of fertilizers through the irrigation system in this manner improves distribution uniformity and increases the probability that the fertilizer will be placed into the root zone. Every effort should be made to prevent movement of fertilizer laden water off-site. Application of fertilizers in irrigation water is sometimes called "fertigation."

## SYSTEM AND FERTILIZER

*Open systems* – These systems include lined and unlined open ditches and gated pipes that are used for furrow and flood irrigation. All forms of liquid and dry fertilizers or amendments can be applied through open systems, though the effectiveness may not be the same for all materials.

Liquid soil amendments, such as ammonium polysulfide, can be applied directly into open systems through metering applicators. Gypsum can be added to the water through weir boxes by diverting part of the stream through a pile of gypsum. About 1 ton of high-grade gypsum can be dissolved in an acre-foot of water. Sulfuric acid may be applied as a soil amendment; however, care must be taken as it is highly corrosive and will attack concrete liners and metal gates in the irrigation system. In these situations, care must be taken to prevent dropping the pH of the irrigation water below pH 4.5.

Equipment for applying dry fertilizer materials through open systems may be inefficient or cumbersome. It may consist of a field worker standing by the irrigation canal pouring a coffee can of fertilizer into the water every so often. It can also be an elaborate platform alongside or straddling the canal, complete with a hopper and a power- or water-driven mechanism to meter fertilizer into the water. Liquid fertilizers are easiest to inject into irrigation systems.

Application of fertilizer solutions through open systems requires less sophisticated equipment than that needed for applying anhydrous ammonia. Tanks are set up by the weir box or check drops, or alongside the irrigation ditch. The tank is connected to a float box that meters the solution into the water. The float box is usually made of hard rubber about the size of an automobile battery casing. The valve is activated by a float.

Equipment necessary for injecting anhydrous ammonia into open irrigation systems consists of a liquid-out valve, a tank, a quick-connect coupling, a line strainer, a flow control pressure regulator, a back-flow check valve, an orifice, a length of hose and a spreader tube at the point where the ammonia enters the water.

Sometimes anhydrous ammonia is metered into the water by means of a fixed orifice flow control. Regardless of the metering system selected, the proper size orifice to be used is determined from the pressure regulator and orifice charts provided by the manufacturer. When injecting anhydrous ammonia into the irrigation ditch, put the tank about 50 feet upstream from the outlets into the field to allow the ammonia to mix well with the water. A practical means of minimizing volatile losses of water-run ammonia, particularly under hot desert conditions, is to maintain ammonia-nitrogen rates at or below 20 pounds of nitrogen per acre-inch of water.

*Closed systems* – These systems include high pressure center pivot, linear and solid-set sprinkler systems as well as low pressure drip, mist and micro-sprinkler irrigation systems. Sprinkler systems usually have all metal plumbing, while the others employ a considerable amount of plastic tubing and fittings.

Not all dry and liquid fertilizers are suitable for application in closed systems. Due to relatively large orifice sizes and high flow rates the high pressure sprinkler systems require less refinement in fertilizer formulation and application technique. In these systems, most liquid fertilizers may be used with the exception of strong acids, aqua ammonia and anhydrous ammonia. Because of the corrosive nature of acid fertilizers and amendments, their use should be reserved for systems constructed out of corrosion-resistant materials. Aqua ammonia and anhydrous ammonia are discouraged in closed systems because of precipitates which may be formed if the irrigation water has a high calcium content. Crop injury may result from sprinkler application of dilute ammonia solutions to plant foliage. In addition, significant volatility losses of the ammonia may occur.

Filters are advised when applying fertilizers through closed irrigation systems, particularly in low volume microirrigation systems. Low operating pressures, non-turbulent flow and small orifice size make these systems more susceptible to plugging than high volume sprinkler systems. To reduce the opportunity for plugging, fertilizers should be introduced into the system at a point well ahead of the filters. Small in-line filters may also be installed in the tubing that connects the field storage tank with the injection apparatus. This technique removes solid material which may be present in the fertilizer tank or transfer truck. Only clean base stock solutions and uncoated dry fertilizers should be used in formulations for low volume micro-irrigation systems.

Various types and models of pumps are available for introducing fertilizer materials into closed irrigation systems. Metering pumps inject the fertilizer into the main irrigation line under pressure. Metering pumps inject fertilizer at a constant rate with respect to water flow in the irrigation system. Proportioner pumps possess flow monitoring devices which adjust fertilizer injection rates based on flow rates so that a constant fertilizer concentration is achieved. Both metering or proportioner pumps may be powered by gas or electric motors, or by water driven pumps which derive their energy from the main irrigation pump.

The pressure developed by these pumps must be greater than the pressure within the closed irrigation system in order for injection to occur. Venturi-type injectors introduce fertilizers by creating a vacuum which draws the fertilizer into the system rather than creating a pressure which forces material into a pressurized system. Venturi-type injectors are very simple devices that involve no moving parts. A short length of rigid tubing is attached to the main irrigation line at two points separated by several feet. The tubing is constricted for a short distance which increases the velocity of water as it passes through this constriction. Connector hose from the fertilizer tank is attached at the point of constriction. A pressure drop accompanies this increase in velocity. A 20 percent pressure differential is sufficient to initiate a negative pressure (vacuum) which draws liquid fertilizer from the connector hose.

Special tanks are available for dissolving dry fertilizers for application through closed systems. Newly developed machines are capable of injecting dissolved gypsum directly into flood, sprinkler or micro-irrigation systems. The injection system consists of a fiberglass tank housing agitation and recirculation devices. Finely ground gypsum is added to the tank, mixed with irrigation water diverted through the tank, and returned to the main irrigation line enriched with dissolved gypsum (calcium sulfate). Users have found these systems to be a simple way to introduce dry fertilizers in closed irrigation systems. Filters are provided to remove insoluble components.

For any system, it is important that the injector be of adequate size to provide the quantity of fertilizer which must be applied based on unit of flow, area, or set. Back-flow prevention devices should be used to prevent fertilizer materials from being accidentally introduced into the water source.

As a special precaution in micro-irrigation systems, thoroughly flush the system with clean water after injection of fertilizers. This effectively prevents accidental mixing of incompatible materials within the irrigation system.

*Nitrogen* – Nearly all forms of nitrogen can be used in closed irrigation systems. As mentioned above, aqua or anhydrous ammonia are usually not recommended. Solutions derived from dry fertilizers are commonly used. However, materials with insoluble coatings should be avoided or, at the very least, thoroughly filtered before delivery to the field.

*Phosphorus* – Use of phosphorus fertilizers in closed irrigation systems, particularly low volume micro-irrigation systems, presents special challenges. In irrigation water having appreciable calcium content (greater than 50 ppm Ca), formation of insoluble calcium phosphates may occur. The higher the calcium concentration, the faster the plugging rate. This may result in reduced phosphorus availability and/or plugging of the irrigation system. Plugging from calcium phosphates is usually not a significant concern in high volume, high flow rate irrigation systems.

Plugging potential from calcium phosphate can be influenced by the source of phosphorus used. Research conducted by TVA has shown that ortho phosphate can be maintained at much higher concentrations in high calcium water than can polyphosphate. The use of furnace grade ortho phosphoric acid is generally recommended for microirrigation systems where precipitation is of concern. Unfortunately, this source of phosphorus is expensive and not readily available in the agricultural market. As an alternative to furnace grade ortho phosphoric acid, injection of sulfuric acid or urea sulfuric acid during or after injection of polyphosphate solutions has been used to prevent or dissolve calcium phosphate precipitates. Achieving a pH of less than 5.0 during injection of phosphorus fertilizers greatly reduces the risk of forming precipitates. Reducing the pH to 3.0 for a short period of time after application of phosphorus will dissolve precipitates present within the system. Prolonged exposure to acidic water may damage aluminum, brass or other sensitive components.

*Potassium* – Potassium nitrate, sulfate, or chloride have all been used successfully in closed irrigation systems. Material costs, crop nutritional requirements and chloride tolerance determine the appropriate source to be used. These materials are typically predissolved prior to use. Potassium solutions which will be used in micro-irrigation systems should be filtered to remove solid contaminants. Due to the low solubility of potassium sulfate, and the cost associated with shipping dilute solutions, some growers prefer to hand apply dry potassium sulfate at the base of trees or vines and allow the irrigation water to slowly dissolve the material. Application of potassium in irrigation water has been shown to be very effective at providing crops with potassium during the growing season. Dissolved potassium is carried into the soil by mass flow with the irrigation water. Vertical distribution in the soil profile to depths as great as 2-3 feet has been measured. Research on prune trees in northern California showed that potassium applied in the irrigation water was superior in terms of crop response to twice the amount shanked into the soil at the drip line of the tree. Potassium applied by this technique remains available for a short period of time before equilibrium between soluble, exchangeable, and fixed forms is restored. This technique has generated promising results in tomatoes, cotton, potatoes, and other crops which have a high seasonal requirement for potassium.

*Micronutrients* – Solutions containing various forms of micronutrients may be used in closed irrigation systems. The solubility of the material being used is usually more important than the form. Insoluble components should be filtered prior to introduction in closed systems.


## WATER QUALITY AND FERTILIZERS

When the irrigation water has a high calcium content, the addition of anhydrous ammonia or liquid ammonium polyphosphate fertilizers may increase the sodium hazard of the irrigation water (see Chapter 2). Anhydrous ammonia causes the calcium

Figure 7-4. Liquid fertilizer containing polyphosphate being added to irrigation water high in dissolved calcium. Immediate formation of calcium phosphate precipitate can lead to plugging of micro-irrigation systems.

in the water to precipitate as solid lime particles, which increases the SAR of the irrigation water. This allows dissolved sodium to attach itself to the soil particles, thus reducing the water intake capacity of the soil. This effect may be reversed with subsequent irrigations during which no anhydrous ammonia is applied. The same problem can occur with ammonium polyphosphate fertilizers. However, since they are applied at rates considerably lower than those of anhydrous ammonia, the adverse effect may not be as serious. Growers should have their irrigation water analyzed to determine whether or not these fertilizers can be applied in the water.

## FOLIAR APPLICATIONS

A portion of the crop's nutritional needs may be met by

Figure 7-5. Plugged drip emitter with cover removed to show precipitate formed in the flow channels.

directly applying fertilizer solutions to the foliage. Foliar nutrition has established itself as a useful technique in the nutritional management of many crops. This technique of fertilizing crops has several potential benefits, including:

1. Supplying nutrients during periods of peak demand when an immediate response is desired.

2. Supplying certain nutrients, such as zinc, when soil or crop conditions are not conducive to root uptake.

3. Precise timing of nutrients related to quality characteristics to the crop being grown.

4. Reducing nitrate leaching in certain cropping systems.

5. Providing a source of nutrients to temporarily satisfy crop demand until a soil application can be made.

For certain situations, such as micronutrients in tree crops, foliar nutrition may offer the most economical and reliable method for correcting and/or preventing deficiencies. For most crops, only a portion of the primary and secondary nutrient requirements can be supplied through the foliage. As a general rule, foliar nutrition programs should supplement, rather than replace, sound soil fertilizer programs. Growers interested in foliar feeding should seek the counsel of farm advisors, extension specialists or industry agronomists regarding the crop to be treated, nutrients and sources that can be applied, rates and methods of application.

Foliar feeding may be achieved by aerial or ground spray applications, or by introducing fertilizers into overhead sprinkler systems. Sprays are usually preferred to overhead irrigation application because of more uniform distribution and less likelihood that the nutrients will be washed off before absorption occurs.

Ground spray equipment used for foliar feeding is usually of the high-pressure, low-volume type, designed to distribute the spray materials uniformly on the foliage and keep water volume to a minimum. The spray may be applied through single-or multiple-nozzle hand guns; multiple-nozzle booms; or by multiple-nozzle, oscillating or stationary cyclone-type orchard sprayers. The response on crops may be affected by droplet size; therefore, the spray used must be regulated by adjusting pressure and selecting the proper nozzles and discs. The concentration of a nutrient in a foliar solution varies widely with respect to the nutrient in question and the crop.

Aerial or ground sprayers usually employ the same equipment used for pesticide applications. In some cases, fertilizers and pesticides may be combined with the same spray. See the following section for more information on fertilizer-pesticide mixtures.

Not all crops respond equally to foliar nutrition. Depending on the soil, crop, and environmental conditions, foliar nutrition may or may not be a suitable method of applying fertilizers. Specific elements differ in their rate of uptake through foliage and the degree of mobility within the plant once absorbed (see Table 7-2). Furthermore, the chemical form of the nutrient (e.g. urea vs. ammonium) may affect responses to foliar nutrition. In regards to micronutrients, the form applied (e.g. chelate vs. sulfate) may impact crop response.

**Table 7–2**
**Generalized Absorption and Mobility Rankings**
**for Foliar Applied Nutrients**

| Absorption | Mobility |
|---|---|
| **Rapid** | **Mobile** |
| Urea Nitrogen | Urea Nitrogen |
| Potassium | Potassium |
| Zinc | Phosphorus |
| | Sulfate |
| **Moderate** | |
| Calcium | **Partially Mobile** |
| Sulfate | Zinc |
| Phosphorus | Copper |
| Manganese | Manganese |
| Boron | Molybdenum |
| | Boron |
| **Slow** | |
| Magnesium | **Immobile** |
| Copper | Iron |
| Iron | Calcium |
| Molybdenum | Magnesium |

Factors which may improve the effectiveness of foliar nutrient sprays include:

1. Application during early morning or evening hours.

2. Application when temperatures are less than 85°F.

3. Relative humidity greater than 70 percent.

4. Inclusion of a high quality spreader adjuvant.

5. Wind speed less than 5 mph.

6. Application to young, actively growing tissue as compared to older, hardened-off tissue.

7. Expanding buds in perennial woody crops effectively absorb topically applied nutrients.

With respect to urea, growers are usually advised to use material with a low biuret content. Biuret is a compound formed during the manufacture of urea. It may be toxic to some plants, particularly when applied to the foliage of perennial crops. Annual crops tend to be more tolerant of biuret.

Examples of successful foliar nutrient programs are listed below. This list is not to be considered inclusive of all known successes with foliar nutrient applications, nor is it meant to imply that success with a program listed is assured.

1. Urea on citrus for increased yield.

2. Urea on small grains for yield and protein enhancement.

3. Phosphorus on potatoes for increased yield.

4. Potassium on cotton for increased yield and reduced disease severity.

5. Potassium on prune for increased yield and reduced limb die-back.

6. Calcium on apples for reduced bitter bit.

7. Zinc or manganese on deciduous or citrus crops to correct nutrient deficiencies.

8. Boron on deciduous fruit and nut crops for improved fruit set and yield.

# CALIBRATION OF
# APPLICATION EQUIPMENT

The rate-of-delivery settings on fertilizer applicators may be predetermined by the equipment manufacturer. Charts showing the settings or orifice sizes for rates of application according to the kinds of fertilizer are usually affixed to the equipment or are given in the operator's manual. The local fertilizer dealer should be consulted for further information.

# FERTILIZER-PESTICIDE
# MIXTURES

Fertilizer-pesticide mixtures are used to fertilize crops and control insects, diseases, nematodes, and weeds. There are more difficulties in making dry fertilizer-pesticide mixtures than there are in formulating liquid fertilizer-pesticide mixtures. For example, when making a dry fertilizer-pesticide mix, the fertilizer should be impregnated with the pesticide to avoid segregation. Even after successful impregnation, the mixture must be applied within a short time or the pesticide may deteriorate. Storing the mixture in airtight containers may retard deterioration, but the cost of the containers and handling may erode any economic gains. A uniform mixture of dry materials with pesticides is difficult to obtain because of the small amount of pesticide added to a large amount of fertilizer.

Pesticides have been successfully mixed and applied with both liquid mix fertilizers and liquid nitrogen solutions. Growers are encouraged to consult with their farm advisors, extension specialists, industrial agronomists, and entomologists for complete details regarding compatibility, stability, and use of fertilizer-pesticide mixtures in crop production.

Even though compatibility and stability factors may present no problems, the nature of the job to be accomplished and the grower's crop management practices may be more influential in making the decision to use fertilizer-pesticide mixtures. For example, it would not be wise to apply a herbicide with a phosphate fertilizer when the herbicide must be lightly incorporated, since the phosphate fertilizer should be more deeply incorporated for best efficiency.

Apart from the agronomic standpoint, there is the aspect of regulatory problems with the use of fertilizer-pesticide mixtures. Legislation, for example, provides that only qualified licensed individuals can formulate, sell, recommend, and apply pesticides. Once a pesticide is added to a fertilizer, the mixture takes on a completely different set of legal ramifications regarding its use. It would be advisable for growers to consult with qualified and licensed technologists before mixing and/or applying fertilizer-pesticide mixtures.

# EMERGING TECHNOLOGY

Innovative approaches to efficient fertilizer application are constantly being sought. Increasing focus on minimizing nitrate movement off-site, and expanded conservation tillage acres have been the driving forces behind two new technologies in fertilizer application. Point injection fertilizer application and variable application rate technologies are being researched, developed, and promoted nationwide.

## POINT INJECTION
## FERTILIZER APPLICATION

Point injection fertilizer application (PIFA) of fertilizer solutions is a sub-surface placement technique that creates discrete zones of nutrients in the crop root zone. Spoke wheel, single probe, and high pressure applicators have been field tested over the past decade. The basic design of a spoke wheel applicator

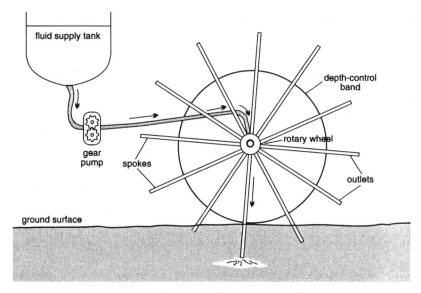

Figure 7-6. Schematic of the spoke wheel PIFA.

consists of a tractor pulled wheel device connected by tubing to the liquid fertilizer supply tank (Figure 7-6). A rotary valve located in the wheel hub accurately meters fluid fertilizer into a ported axle so that only the outlet which is positioned downward receives fertilizer. The fertilizer thus flows from the spokes as the wheel rotates. Spoke wheel units are mounted on a tool bar to accommodate variable row spacings. Fertilizer is typically delivered to the soil at 8-inch spacing and to a depth of 3-4 inches.

Though initial field research was conducted in the mid-west, testing on a wider range of crops, geographic areas, and climatic conditions is producing encouraging agronomic performance data. A variety of data available suggest that the following agronomic benefits can be anticipated through the spoke wheel PIFA.

1. Energy savings. A spoke wheel PIFA may have only one-third of the power requirement of knife injection systems.

2. Facilitates post plant sub-surface fertilizer placement without root pruning commonly associated with conventional knife applications.

3. Minimal soil and residue disturbance.

4. Improved yields obtained by increased efficiencies of split timing and placement, sometimes with fertilizer rate reductions.

5. Reduced leaching, run-off, and denitrification losses leading to improved nitrogen use efficiencies.

## VARIABLE RATE TECHNOLOGY

Variable rate technology (VRT) refers to changing fertilizer ingredients and application rate "on the go" based on differences in soil fertility levels within fields. The VRT process involves a number of different components working in concert to achieve highly efficient fertilizer application. VRT has emerged as perhaps the most technologically advanced approach to managing crop nutrition. The basic concept uses information from individual units or cells within a field to govern fertilizer materials and amounts to achieve desired rates.

The cornerstones for VRT are systematic soil sampling repeated at prescribed intervals within fields, digitized mapping of soil fertility levels, computer software which allows the application equipment to "read" the digitized map, and sophisticated application equipment that can change fertilizer materials and rates on the go. Spatial referencing within fields may be as simple as computer aided monitoring based on ground speed and a digitized soil map to high-tech linking with the Global Positioning System, a constellation of satellites developed by the U.S. government.

Most VRT for fertilizers involves soil sampling, analysis, and programming of computer software well in advance of the planned fertilizer application. Newer research is being directed towards development of processes that allow for real time VRT. In real time VRT, analysis, decision making, and variable fertilizer application rate all take place simultaneously as the application equipment passes through the field. Though not currently developed to the point of commercialization, real time VRT offers perhaps the single most significant technology affecting fertilizer use since the development of synthetic fertilizers during the 1940's.

## SUPPLEMENTARY READING

1. *Advances in Production and Utilization of Quality Cotton: Principles and Practices.* F. C. Elliot, M. Hoover and W. K. Porter, Jr., eds. Iowa State University Press. 1968.

2. *Advances in Sugarbeet Production: Principles and Practices.* R. T. Johnson, J. T. Alexander, G. E. Rush and G. R. Hawkes, eds. Iowa State University Press. 1971.

3. *Agricultural Anhydrous Ammonia: Technology and Use.* M. H. McVickar, W. P. Martin, I. E. Miles and H. H. Tucker, eds. Soil Sci. Soc. of America. 1966.

4. *Bulk Blend Quality Control Manual.* The Fertilizer Institute. June 1975.

5. *The Citrus Industry,* Volume III. W. Reuther. University of California. 1973.

6. *Fertilizer Technology and Usage.* M. H. McVickar, G. L. Bridger and L. B. Nelson. Soil Sci. Soc. of America. 1963.

7. *Nitrogen in Agricultural Soils.* F. J. Stevenson, ed. American Society of Agronomy. 1982.

8. *Point Injection Fertilizer Application (PIFA).* J. Julian Smith and Paul S. Belzer. Proceedings 1993 California Plant and Soil Conference.

9. *The Role of Phosphorus in Agriculture.* F. E. Khasawneh,

E. C. Sample and E. J. Kamprath, eds. American Society of Agronomy. 1968.

10. *The Role of Potassium in Agriculture.* V. J. Kilmer, S. E. Younts and N. C. Brady, eds. American Society of Agronomy. 1968.

11. *Soil Fertility and Fertilizers,* Fourth Edition. S. L. Tisdale, W. L. Nelson, and J. D. Beeton. The Macmillan Company. 1985.

12. *Soil Specific Crop Management: A Workshop on Research and Developmental Issues.* American Society of Agronomy. 1992.

13. TVA Fertilizer Bulk Blending Conference. Louisville, Ky. August 1-2, 1973.

14. *Using Commercial Fertilizers,* Fourth Edition. M. H. McVickar and W. M. Walker. The Interstate Printers & Publishers, Inc. 1978.

# Chapter 8

# *Soil and Tissue Testing*

Soil and plant tissue analyses are the best guide to the wise and efficient use of fertilizers and soil amendments. They are a recommended best management practice which can help in the production of high yields and high quality crops while maintaining environmental quality. Useful recommendations based on soil and tissue tests require accurate sampling, analysis and interpretation based on sound research, practical experience, and good judgment. The interpretive guides presented here apply generally throughout the West. These guides should be used only with data obtained from samples collected and analyzed by the procedures specified. The user is advised to contact the local agricultural extension service, experiment station, or qualified industry representatives for recommendations for specific cropping systems.

Each diagnostic technique has both advantages and limitations. Soil and plant analyses do different jobs and should be used in such a way that they support and supplement each other. Soil analyses are most useful in appraising nutrient requirements and in evaluating pH and salt problems. They have the advantage that they can be completed and the information used prior to planting.

Plant analyses are particularly useful for determining the nutritional status of established, deep-rooted perennials, such as alfalfa, trees, and vines, where soil samples of the entire feeding zone are difficult to obtain and interpret. They are also useful for diagnosing the causes of poor growth, in evaluating the effectiveness of fertilizer treatments, in following the nutrient status of plants throughout the growing season, and in managing quality factors.

Satisfactory recommendations based on soil or tissue tests depend upon three factors: representative sampling, accurate analysis, and proper interpretation of the analytical results.

## *SOIL TESTING*

### *SAMPLING*

The first step in soil testing is the collection of a representative sample. The sample must accurately represent the soil in which the crop is being grown. The analytical results obtained from the laboratory can be no better than the sample which is collected from the field. Sample collection is complex and probably the greatest cause of variability in soil testing.

The sample submitted to an analytical laboratory will usually weigh less than one pound, but may represent over 100 million pounds of soil. Fertilizer recommendations costing hundreds or thousands of dollars are based on the results of the laboratory analysis. It is foolish indeed to send a non-representative sample to the laboratory.

Soils can be extremely heterogeneous (Figure 8-1). They vary in both their horizontal and vertical dimensions. Management practices such as levelling, fertilizing, and cropping can further increase this natural heterogeneity. This variability should be recognized and, if possible, taken into account when soil samples are collected and fertilizer applications made. On a practical basis, samples are usually collected to represent areas which may be fertilized and managed separately. A minimum of 20 subsamples should be collected from within this manage-

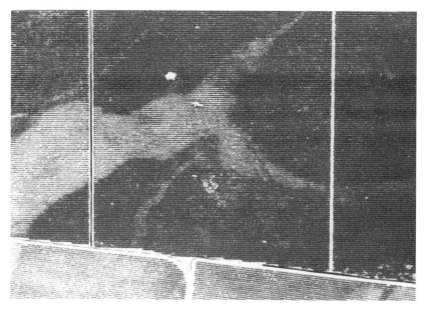

Figure 8-1. Aerial view of a vineyard showing variability.

ment area. The larger the area represented by the sample, the greater the possibility of wide variability within the area.

Soil samples for standard analysis (P, K, Ca, Mg, etc.) should be collected to a depth of one foot. Samples to be analyzed for salts, sodium, and mobile nutrients, such as nitrate ($NO_3^-$), should be collected to the rooting depth of the crop. Samples to be analyzed for $NO_3^-$ should be dried as soon as possible after collection. If immediate drying is not possible, the samples should be frozen until they can be dried or delivered to the laboratory for analysis. Avoid excessive heat when drying samples.

Sample collection within fields which have received banded applications of fertilizer without mixing presents special problems. Collection of a representative sample is extremely difficult from such a field. Great care must be taken when interpreting the results from fields which have received banded fertilizer applications. Consult your laboratory or extension specialist for specific instructions.

Soil samples can also prove quite helpful when attempting to diagnose field problems. Paired samples should be collected from adjacent normal and affected areas and the results compared.

Before collecting and submitting soil samples for analysis, be sure to obtain specific information from the laboratory conducting the analysis. The laboratory will also have the proper sample bags and information sheets. Be sure to keep detailed records on the areas sampled, the fertilizers used, crops grown and residue management, yields, and other pertinent production information for future reference.

## SAMPLE ANALYSIS

Soils may be analyzed for a wide range of chemical and physical parameters. A typical chemical analysis of soil could include pH, free lime, N, P, K, Ca, Mg, S, Na, C.E.C., micro-

Figure 8-2. Modern soil and plant analysis laboratories use sophisticated automated equipment.

nutrients, organic matter, and salts. However, not all samples need to be analyzed for all these factors.

Soil pH is measured using a glass electrode in a 1:1 or 1:2 soil:water paste. Both nitrate-N and ammonic-N can be determined to assess the nitrogen fertility status of a soil. Phosphorus is usually measured by extracting the soil with $NaHCO_3$ for neutral to calcareous soils and with a mild acid for acidic soils. Exchangeable cations are usually determined following extraction with ammonium acetate. Micronutrients are extracted using DTPA. Detailed methods of analysis for these determinations are available from several sources (see Supplementary Reading).

## INTERPRETATION OF
## ANALYTICAL RESULTS

Soil tests do not measure "plant available" nutrients or "pounds of available nutrient" (except N and S). Soil testing actually measures an index of the amount of plant nutrient in the soil and this is then correlated with the likelihood of a response to fertilizer through soil test calibration.

The results from laboratory analysis must be interpreted before meaningful fertilizer or other recommendations can be made. The interpretation of the results from soil testing must take into consideration additional factors such as crop species and variety, crop rotation effects, environmental considerations, management capability of the grower, and other factors.

A typical yield response curve relating soil test values to crop yield is shown in Figure 8-3. Crop yields are very low at low soil test values. As the soil test level for a given nutrient increases, the crop yield increases until, at some point, there is no additional yield increase. This point is termed the soil test critical level. Sites with this soil test value or higher will not respond to application of that nutrient. Conversely, sites with soil test values below this point will respond to application of that nutrient.

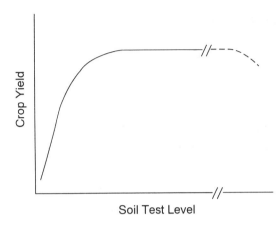

Figure 8-3. Typical soil test yield response curve.

Some nutrients such as Zn, Mn, and B may become toxic at extremely high levels and actually reduce yields.

Some general guidelines for interpreting soil tests for P, K, and Zn are shown in Table 8-1. Keep in mind that these are

### Table 8–1
### Generalized Soil Test Interpretive Guide

| Most Field Crops | NaHCO$_3$–P[1] | Exchange-able K | DTPA Zn |
|---|---|---|---|
| Probability of Response | - - - - - - - - - - *(ppm)* - - - - - - - - - - | | |
| High | <10 | <100 | <0.5 |
| Medium | 10–15 | 100–150 | 0.5–1.0 |
| Low | >15 | >150 | >1.0 |
| **Vegetable Crops and Potatoes** | **NaHCO$_3$–P[2]** | **Exchange-able K[3]** | **DTPA Zn** |
| Probability of Response | - - - - - - - - - - *(ppm)* - - - - - - - - - - | | |
| High | <15 | <100 | <0.5 |
| Medium | 15–25 | 100–200 | 0.5–1.0 |
| Low | >25 | >200 | >1.0 |

[1]For cool season crops, increase these values 30%.

[2]Probability of response is greater on high lime soils.

[3]Exchangeable K values 50% to 100% higher are suggested for Washington and Oregon.

general guidelines; consult with local experts for specific recommendations.

## ADDITIONAL SOIL TESTS

In addition to the standard soil tests discussed above, there are new and specialized soil tests which may offer more information about certain soils. For example, the use of ion exchange resins is offering the possibility of increased ease of analysis with improved accuracy. These exchange resins are used in various ways but they are conceptually similar. The resin is placed in the soil for a period of time during which the various nutrient ions are adsorbed onto the resin. The resin is then removed from the soil and the concentration of the ions is measured.

A soil test has been developed which measures the rate at which K is released from the exchangeable fraction to the soil solution fraction. This test has proven quite helpful in identifying sites at which the K release rate does not meet the crop demand for K. These sites often respond to K fertilizer even though the standard K soil test indicates adequate K.

Technology is now becoming available which permits the variable application of fertilizer in response to changing soil test levels within a field. This concept has been called "variable application rate technology" (VART) or "site-specific fertilizing." More information is given in Chapter 7, "Methods of Applying Fertilizer." The system is based on grid sampling the field, storing the results in computer maps, and then varying the rate of fertilization to match the varying soil test levels within the field.

The Pre-Sidedress Nitrate Test (PSNT) is being used in the corn belt to refine sidedress N application rates. It is essentially a standard nitrate soil test conducted immediately prior to sidedressing. The test has proven to be of some value, but much more work needs to be done. Contact your local experts for more information on this test.

# PLANT ANALYSIS

## SAMPLING

Collecting a representative plant sample is essential if reliable results are to be obtained. When collecting plant samples, the crop and its stage of growth, the particular plant part to collect, and the form of the nutrient to be measured must all be considered. The nutrient content can vary widely during the growing season and it can also vary widely between different plant parts. The values obtained can be compared to known standards only if the sample has been properly collected and documented.

Plant analysis can also prove quite helpful when attempting to diagnose field problems. Paired samples should be collected from adjacent normal and affected areas and the results compared.

Consult the following tables and figures to determine the appropriate stage of growth, plant part, and form of the nutrient to be analyzed. If questions remain, be sure to contact the laboratory conducting the analysis for guidelines.

## SAMPLE ANALYSIS

Plant tissue analysis can be used to identify inadequate, marginal, normal, or excessive concentrations of many nutrients. Plant samples may be analyzed for nitrate-N ($NO_3$-N), total N, acetic acid-soluble phosphate ($PO_4$-P), total P, sulfate-S ($SO_4$-S), total S, K, Ca, Mg, and micronutrients. Not all samples need to be analyzed for all these nutrients. Detailed methods of

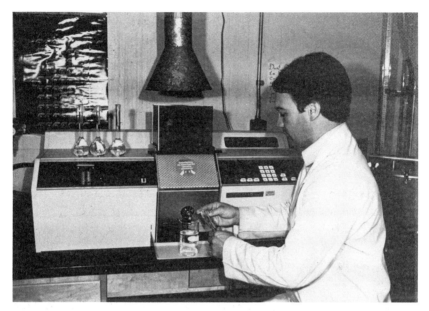

Figure 8-4. Measuring nutrient concentration with an atomic absorption spectrophotometer.

analysis for these determinations are available from several sources (see Supplemental Reading).

## *INTERPRETATION OF*
## *ANALYTICAL RESULTS*

Plant analysis results are the actual concentration of nutrients found in the sample. It is not an "index" like a soil test. A typical response curve relating nutrient concentration to crop yield is shown in Figure 8-5. Crop yields are very low at low nutrient concentrations. As the nutrient concentration in the crop increases, the yield increases until, at some point, there is no additional yield increase. This point is termed the plant analysis critical level. Plants with this or greater concentrations of the particular nutrient will not respond to additional nutrient. Conversely, plants with this or lesser concentrations of nutrient

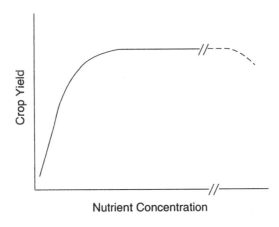

Figure 8-5. Typical plant analysis yield response curve.

will respond to additional nutrient. Extremely high concentrations of some nutrients may result in yield reductions due to toxicity or nutrient imbalance.

The concentration of nutrients found in the plant sample can be compared to published standards and guidelines. The guidelines may need to be modified and adjusted for differences in environmental factors, other stresses on the plant, and so forth.

The nutritional status of many perennial crops such as tree fruits, grapes, and alfalfa, may be evaluated from samples collected at a single specific growth stage. For annual crops, such as potatoes, vegetables, and sugarbeets, it is better to collect several samples during the growing season so trends in the nutritional status can be assessed and, if necessary, corrected as soon as possible.

Guidelines for the interpretation of tissue tests for many western crops are given in Tables 8-2 through 8-8. Nutrient levels are given in relation to the probability of a response from the application of fertilizers supplying the nutrient. Tissue test values below the critical nutrient range (CNR) indicate a high probability of response; values above the CNR indicate a low probability of response. A response may or may not be obtained when tissue test levels are within the CNR depending on production levels and other factors. Concentrations greater than the "excess" levels may be associated with depressed yields or crop quality. It is very important that samples of the same plant part, taken at the same physiological age, be analyzed if the values presented here are to be used as a guideline.

The influence of leaf levels of nitrogen and potassium on yield and quality factors of oranges is presented in Figure 8-6 and 8-7. Increasing the level of a given nutrient in the tree influences some factors favorably and others unfavorably. In

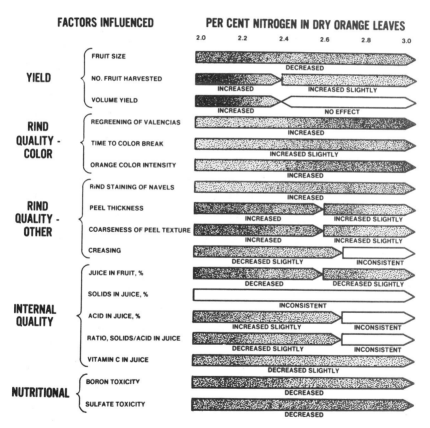

Figure 8-6. Influence of the percentage of N in 5- to 7-month-old bloom-cycle leaves from nonfruiting shoots upon yield, and rind and fruit quality of oranges (adapted from Embleton, *et al.*, 1973).

most years, returns are maximum when the nitrogen and potassium concentrations are in the optimum ranges. In other years, and in some orchards, adjustments around the optimum ranges may be worthwhile.

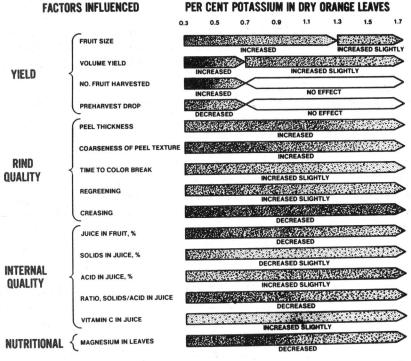

Figure 8-7. Influence of the percentage of K in 5- to 7-month-old bloom-cycle leaves from nonfruiting shoots upon yield, and rind and fruit quality of oranges (adapted from Embleton, Jones and Platt in *Soil and Plant-Tissue Testing in California*, 1978).

## Table 8–2
## Plant Tissue Analysis Guidelines for Many Western Crops

| Crop/Time or Growth Stage of Sampling | Plant Part | Nutrient | Critical Nutrient Range |
|---|---|---|---|
| **Alfalfa** | | | |
| 1/10 bloom | Tops, upper | P | 0.2–0.3% |
| | Mid-stem | $PO_4$-P | 800–1,000 ppm |
| | Tops, upper 1/3 | K | 2.0–2.5% |
| | Tops upper 3/4 | S | N/S ratio = 11 |
| | Tops, upper 2/3 | S | 0.19–0.21% |
| | Whole tops | $SO_4$-S | 0.08–0.10% |
| | Tops, upper 2/3 | B | 20–30 ppm |
| | Top 1/2 of shoots | Zn | 20–25 ppm |
| | Tops, upper 2/3 | Mo | 0.1–0.2 ppm |
| **Asparagus** | | | |
| Midgrowth of fern | 4" tip section of new fern branch | $NO_3$-N | 100–500 ppm |
| | | $PO_4$-P | 800–1,600 ppm |
| | | K | 1–3% |
| **Barley** | | | |
| 3-4 leaf | Below ground stem | $NO_3$-N | 600–700 ppm |
| After heading | Leaves, second node | Zn | 20–25 ppm |
| Early tillering | Top half w/leaves | Mn | 25–30 ppm |
| Tillering | Leaves | Cu | < −4 ppm |
| Tillering | Leaves | B | < −5 ppm |
| **Bean, bush snap** | | | |
| Midgrowth | Petiole of fourth leaf from growing tip | $NO_3$-N | 2,000–4,000 ppm |
| | | $PO_4$-P | 1,000–3,000 ppm |
| | | K | 3–5% |
| Early bloom | Petiole of fourth leaf from growing tip | $NO_3$-N | 1,000–2,000 ppm |
| | | $PO_4$-P | 800–2,000 ppm |
| | | K | 2–4% |
| **Bean, dry** | | | |
| June | Recently expanded trifoliate leaves | N | 4–5% |
| June–July | | P | 0.25–0.35% |
| Pre-bloom | Whole tops | K | 2.6–3.2% |
| Bud stage | Whole tops | Zn | 24–30 ppm |

*(Continued)*

## Table 8–2 (Continued)

| Crop/Time or Growth Stage of Sampling | Plant Part | Nutrient | Critical Nutrient Range |
|---|---|---|---|
| **Broccoli** | | | |
| Midgrowth | Mid-rib of young, mature leaf | $NO_3$-N | 7,000–10,000 ppm |
| | | $PO_4$-P | 2,500–5,000 ppm |
| | | K | 3–5% |
| First buds | Mid-rib of young, mature leaf | $NO_3$-N | 5,000–9,000 ppm |
| | | $PO_4$-P | 2,000–4,000 ppm |
| | | K | 2–4% |
| **Brussels sprouts** | | | |
| Midgrowth | Mid-rib of young, mature leaf | $NO_3$-N | 5,000–9,000 ppm |
| | | $PO_4$-P | 2,000–3,500 ppm |
| | | K | 3–5% |
| Late growth | Mid-rib of young, mature leaf | $NO_3$-N | 2,000–4,000 ppm |
| | | $PO_4$-P | 1,000–3,000 ppm |
| | | K | 2–4% |
| **Cabbage** | | | |
| At heading | Mid-rib of wrapper leaf | $NO_3$-N | 5,000–9,000 ppm |
| | | $PO_4$-P | 2,500–3,500 ppm |
| | | K | 2–4% |
| **Cantaloupe** | | | |
| Early growth (short runners) | Petiole of sixth leaf from growing tip (all growth stages) | $NO_3$-N | 8,000–12,000 ppm |
| | | $PO_4$-P | 2,000–4,000 ppm |
| | | K | 4–6% |
| Early fruit | | $NO_3$-N | 5,000–9,000 ppm |
| | | $PO_4$-P | 1,500–2,500 ppm |
| | | K | 3–5% |
| First mature fruit | | $NO_3$-N | 2,000–4,000 ppm |
| | | $PO_4$-P | 1,000–2,000 ppm |
| | | K | 2–4% |
| **Carrot** | | | |
| Midgrowth | Petiole of young, mature leaf | $NO_3$-N | 5,000–10,000 ppm |
| | | $PO_4$-P | 2,000–4,000 ppm |
| | | K | 4–6% |

*(Continued)*

## Table 8–2 (Continued)

| Crop/Time or Growth Stage of Sampling | Plant Part | Nutrient | Critical Nutrient Range |
|---|---|---|---|
| **Cauliflower** | | | |
| Buttoning | Mid-rib of young, mature leaf | $NO_3$-N | 5,000–9,000 ppm |
| | | $PO_4$-P | 2,500–3,500 ppm |
| | | K | 2–4% |
| **Celery** | | | |
| Midgrowth | Petiole of newest fully elongated leaf (all growth stages) | $NO_3$-N | 5,000–9,000 ppm |
| | | $PO_4$-P | 2,000–4,000 ppm |
| | | K | 4–7% |
| Near maturity | | $NO_3$-N | 4,000–6,000 ppm |
| | | $PO_4$-P | 2,000–4,000 ppm |
| | | K | 3–5% |
| **Clover, red** | | | |
| 1/10 bloom | Tops, upper 1/2 | P | < −0.25% |
| | Whole tops | B | < −20 ppm |
| **Clover, subterranean** | | | |
| Late bloom | Leaflets and petiole | P | < −.25%[1] |
| | | K | 0.7–1.0% |
| | | S | 0.15–0.20% |
| | | B | < −20 ppm |
| **Clover, white** | | | |
| 1/2 bloom | Tops, upper 1/2 | P | 0.30–0.35% |
| **Corn, field** | | | |
| Early silk | Early leaf | P | 0.3–0.4% |
| Early silk | Early leaf | $PO_4$-P | 1,200–1,500 ppm |
| Silking | Whole leaf opposite & below ear | K | 2.0–2.5% |
| Pollination | 6th leaf from bottom | Zn | 20–25 ppm |
| Silking | Fully mature leaves | B | 20 ppm |

(Continued)

## Table 8–2 (Continued)

| Crop/Time or Growth Stage of Sampling | Plant Part | Nutrient | Critical Nutrient Range |
|---|---|---|---|
| **Cucumber (pickling)** | | | |
| Early fruit-set | Petiole of sixth leaf from growing tip | $NO_3$-N $PO_4$-P K | 5,000–9,000 ppm 1,500–2,500 ppm 3–5% |
| **Grass, orchard (pasture)** | | | |
| 12 in. height | Tops | N | 3.5–4.0% |
| **Grass, orchard and perennial rye** | | | |
| 6 in. height | Tops | P K | 0.35–0.40% 1.8–2.2% |
| **Grass, orchard and tall fescue** | | | |
| Harvest early bloom | Tops | S | 0.10–0.15% |
| **Hops** | | | |
| Early bloom | 1st leaf sidearms | P Zn | 0.18–0.25% 12–20 ppm |
| **Lettuce** | | | |
| At heading | Mid-rib of wrapper leaf (all growth stages) | $NO_3$-N $PO_4$-P K | 4,000–8,000 ppm[2] 2,000–4,000 ppm 2–4% |
| At harvest | | $NO_3$-N $PO_4$-P K | 3,000–6,000 ppm[2] 1,500–2,500 ppm 1.5–2.5% |
| **Mint** | | | |
| Early bloom | 1st unbranched stem from top | N P K Zn | 2.0–3.0% 0.3–0.4% 2.0–3.0% 20–25 ppm |

*(Continued)*

## Table 8–2 (Continued)

| Crop/Time or Growth Stage of Sampling | Plant Part | Nutrient | Critical Nutrient Range |
|---|---|---|---|
| **Oats** | | | |
| 3-4 leaf | Below ground stems | $NO_3$-N | 600–700 ppm |
| Early tiller | Flag leaf | Mn | 25–30 ppm |
| **Peas** | | | |
| Late bloom | Tops | P | 0.20–0.25% |
| 4–8 node | Leaves 3rd node from top | $PO_4$-P | 800–1,000 ppm |
| Mid growth | Leaf | N | 4.8–5.3% |
| Mid growth | Leaf | K | 2.2–3.2% |
| **Pepper, chile** | | | |
| Early growth | Petiole of young, mature leaf (all growth stages) | $NO_3$-N | 5,000–7,000 ppm |
| | | $PO_4$-P | 2,000–3,000 ppm |
| | | K | 4–6% |
| Early fruit-set | | $NO_3$-N | 1,000–2,000 ppm |
| | | $PO_4$-P | 1,500–2,500 ppm |
| | | K | 3–5% |
| **Pepper, sweet** | | | |
| Early growth | Petiole of young, mature leaf (all growth stages) | $NO_3$-N | 8,000–12,000 ppm |
| | | $PO_4$-P | 2,000–4,000 ppm |
| | | K | 4–6% |
| Early fruit-set | | $NO_3$-N | 3,000–5,000 ppm |
| | | $PO_4$-P | 1,500–2,500 ppm |
| | | K | 3–5% |
| **Potatoes** | | | |
| Early tuber set | Recently mature or 4th petiole from top (all growth stages) | $NO_3$-N | 15,000–18,000 ppm |
| Tubers 3/4 in. diam. | | $NO_3$-N | 1.6–2.1% |
| 30 days after 3/4 in. diam. | | $NO_3$-N | 1.2–1.6% |
| 60 days after 3/4 in. diam. | | $NO_3$-N | 0.9–1.2% |

*(Continued)*

## Table 8–2 (Continued)

| Crop/Time or Growth Stage of Sampling | Plant Part | Nutrient | Critical Nutrient Range |
|---|---|---|---|
| **Potatoes (continued)** | | | |
| Early tuber set | | $PO_4$-P | 1,400–1,600 ppm |
| Tubers 3/4 in. diam. | | P | 0.30–0.35% |
| 30 days after 3/4 in. diam. | | P | 0.2–0.3% |
| 60 days after 3/4 in. diam. | | P | 0.10–0.2% |
| Tubers 3/4 in. diam. | | K | 9.5–11.0% |
| 30 days after 3/4 in. diam. | | K | 9.0–10.0% |
| 60 days after 3/4 in. diam. | | K | 8.0–9.0% |
| Early tuber set | | Zn | 10–20 ppm |
| Tubers 3/4 in. diam. | | Mn | 30–35 ppm |
| Early tuber set | | B | 20–30 ppm |
| **Rice** | | | |
| Midtillering | Young, mature leaf — the "y" leaf (all growth stages) | N | < –3.0% |
| | | $PO_4$-P | < –1,000 ppm |
| | | K | 1.2–1.4% |
| Maximum tillering | | N | 2.6–2.8% |
| | | $PO_4$-P | 800–1,000 ppm |
| | | K | 1.0–1.2% |
| Panicle initiation | | N | 2.4–2.6% |
| | | $PO_4$-P | 800–1,000 ppm |
| | | K | 0.8–1.0% |
| **Rose clover** | | | |
| Flowering | Leaves | $PO_4$-P | 1,200–1,500 ppm |
| | | K | 0.7–1.0% |
| | | $SO_4$-S | 130–180 ppm |

*(Continued)*

## Table 8–2 (Continued)

| Crop/Time or Growth Stage of Sampling | Plant Part | Nutrient | Critical Nutrient Range |
|---|---|---|---|
| **Spinach** | | | |
| Midgrowth | Petiole of young, mature leaf | $NO_3$-N | 4,000–8,000 ppm |
| | | $PO_4$-P | 2,000–4,000 ppm |
| | | K | 2–4% |
| **Sweet corn** | | | |
| Tasseling | Mid-rib of first leaf above primary ear | $NO_3$-N | 500–1,500 ppm |
| | | $PO_4$-P | 500–1,000 ppm |
| | | K | 2–4% |
| **Sweet potato** | | | |
| Midgrowth | Petiole of sixth leaf from growing tip | $NO_3$-N | 1,500–3,500 ppm |
| | | $PO_4$-P | 1,000–2,000 ppm |
| | | K | 3–5% |
| **Tomato (canning)** | | | |
| Early bloom | Petiole of fourth leaf from growing tip (all growth stages) | $NO_3$-N | 10,000–13,000 ppm |
| | | $PO_4$-P | 3,000–4,000 ppm |
| | | K | 4–6% |
| Fruit 1" diameter | | $NO_3$-N | 6,000–8,000 ppm |
| | | $PO_4$-P | 2,500–3,500 ppm |
| | | K | 3.5–4.5% |
| First color | | $NO_3$-N | 3,000–5,000 ppm |
| | | $PO_4$-P | 2,000–3,000 ppm |
| | | K | 2–4% |
| **Watermelon** | | | |
| Early fruit | Petiole of sixth leaf from growing tip | $NO_3$-N | 5,000–9,000 ppm |
| | | $PO_4$-P | 1,500–2,500 ppm |
| | | K | 3–5% |
| **Wheat** | | | |
| Boot | Top two leaves | N | 2.3–2.7% |
| Jointing | Total tops | N | 2.5–3.0% |
| 3–4 leaf | Underground stems | $NO_3$-N | 0.25–0.35% |

*(Continued)*

## Table 8–2 (Continued)

| Crop/Time or Growth Stage of Sampling | Plant Part | Nutrient | Critical Nutrient Range |
|---|---|---|---|
| **Wheat (continued)** | | | |
| Joint | 1st 2 inches above ground | $NO_3$-N | 0.08–0.15% |
| Boot | Two top leaves | P | 0.25–0.30% |
| Jointing | Total tops | P | 0.32–0.40% |
| Boot | Whole plant with crown | $PO_4$-P | 0.08–0.10% |
| Early boot | Total tops | K | 1.5–2.0% |
| Boot | Two top leaves | K | 1.5–1.7% |
| Head | Total tops | K | 1.25–1.75% |
| Jointing | Total tops | K | 2.0–2.5% |
| After heading | Leaves, second node | Zn | 20–25 ppm |
| Early tillering | Top half of leaves | Mn | 25–30 ppm |
| Tillering | Leaves | B | < –5 ppm |

[1]The sufficient level of P for heavily grazed subclover may be 50% higher.

[2]Nitrate concentrations 30% higher are suggested for winter-grown lettuce in the desert valleys of Arizona and southern California.

## Table 8–3
## Plant Nutrient Levels in Sugar Beets[1]

| Nutrient | Plant Part | Nutrient Range in Which Deficiency Symptoms May Appear | Critical Concentration[2] |
|---|---|---|---|
|  |  | (ppm) | (ppm) |
| PO$_4$-P | Petiole | 150–400 | 750 |
|  | Blade | 250–700 | — |
| Total K |  |  |  |
| >1.5% Na | Petiole | 2,000–6,000 | 10,000 |
|  | Blade | 3,000–6,000 | 10,000 |
| <1.5% Na | Blade | 4,000–5,000 | 10,000 |
| Total Ca | Petiole | 400–1,000 | 1,000 |
|  | Blade | 1,000–4,000 | 5,000 |
| Total Mg | Petiole | 100–300 | — |
|  | Blade | 250–500 | — |
| SO$_4$-S | Blade | 50–200 | 250 |
| Total Mn | Blade | 4–20 | 10 |
| Total Fe | Blade | 20–55 | 55 |
| Total Zn | Blade | 2–13 | 9 |
| Total B | Blade | 12–40 | 27 |

[1]Adapted from Hills, Salisbery and Ulrich, 1978.

[2]The critical concentration is that nutrient concentration at which growth of a plant is retarded by 10%.

### Table 8–4
### Interpretive Guide for Upland Cotton Petiole Analysis[1]

| Time of Sampling | Desirable or "Safe" Levels | | |
| --- | --- | --- | --- |
| | $NO_3$-N[2] | $PO_4$-P | K |
| | *(ppm)* | *(ppm)* | *(%)* |
| First square | 15,000–18,000 | — | — |
| First bloom | 12,000–18,000 | 1,500–2,000 | 4.0–5.5 |
| Peak bloom | 3,000–7,000 | 1,200–1,500 | 3.0–4.0 |
| First open boll | 1,500–3,500 | 1,000–1,200 | 2.0–3.0 |
| Maturity | Less than 2,000 | 800–1,000 | 1.0–2.0 |

[1]Data are for composition of petioles of the most recent, fully expanded leaf on the main stem (usually the third or forth leaf from the terminal). From data of D. M. Bassett, A. J. MacKenzie and T. C. Tucker.

[2]Pima levels may be somewhat lower.

### Table 8–5
### Interpretive Guide for Thompson Seedless Grape Tissue Analysis[1]

| Nutrient | Deficient | Sufficient | Excess |
| --- | --- | --- | --- |
| $NO_3$-N, ppm | <350 | 600–1,200 | >2,400 |
| P (total), % | <0.15 | 0.20–0.60 | — |
| K (total), % | <1.0 | 1.50–2.50 | >3.0 |
| Mg (total), % | <0.3 | 0.5–0.8 | >1.0 |
| Zn (total), ppm | <15 | 25–50 | — |
| B (total), ppm | <25 | 40–60 | >300[2] |
| Cl (total), % | — | 0.05–0.15 | >0.50 |

[1]Concentrations of nutrients in petioles collected opposite the cluster at full-bloom. Adapted from information supplied by J. A. Cook.

[2]Concentration in leaf-blade tissue.

## Table 8-6
## Leaf Analysis Guide for Fruit and Nut Trees, Pacific Northwest[1]

|  | Nutrient | Level of Nutrient in Leaves[2] | | |
|---|---|---|---|---|
|  |  | Deficient Below | Satisfactory | Excess Above |
| Peach | N | 3.2% | 3.2–3.8% | 3.8% |
| Sweet cherry | N | 2.3% | 2.3–3.0% | 3.0% |
| Apricot, prune | N | 2.0% | 2.0–2.5% | 3.0% |
| Filbert, walnut | N | 2.2% | 2.2–2.8% | 2.8% |
| Pear | N | 1.7% | 1.9–2.5% | 2.5% |
| Apple, pear | K | 0.9% | 1.0–1.5% | — |
| Peach | K | 1.5% | 1.5–3.5% | — |
| Sweet cherry | K | 1.2% | 1.5–3.0% | — |
| Prune | K | 1.3% | 1.7–2.7% | — |
| Filbert, walnut | K | 1.0% | 1.2–? % | — |
| Apple, pear | B | 20 ppm | 30–80 ppm | 100 ppm |
| Stone fruits | B | 20 ppm | 35–80 ppm | 100 ppm |
| All fruits | Zn | 10 ppm | 17–? ppm | — |

[1]Adapted from Oregon State University FG 23, 24, 25, 26, 34 and 35; and Washington State University FG 28f and 28g.

[2]Midshoot leaves sampled between July 15 and August 15.

Table 8–7
Leaf Analysis Guide for Fruit and Nut Trees, California[1]

| | N[2] Adequate | K[3] Deficient | K[3] Adequate | Na[4] Excess | Cl[4] Excess | B Adequate | B Excess |
|---|---|---|---|---|---|---|---|
| | - - - - - - - - - - - (%) - - - - - - - - - - - | | | | | - - - (ppm) - - - | |
| mond | 2.0–2.5 | 1.0 | 1.4 | 0.25 | 0.3 | 30–65 | 85 |
| ple | 2.0–2.4 | 1.0 | 1.2 | — | 0.3 | 25–70 | 100 |
| ricot (ship) | 2.0–2.5 | 2.0 | 2.5 | 0.1 | 0.2 | 20–70 | 90 |
| ricot (can) | 2.5–3.0 | 2.0 | 2.5 | 0.1 | 0.2 | 20–70 | 90 |
| eet cherry | 2.0–3.0 | 0.9 | — | — | — | — | — |
| ] | 2.0–2.5 | 0.7 | 1.0 | — | — | — | 300 |
| ive | 1.5–2.0 | 0.4 | 0.8 | 0.2 | 0.5 | 20–150 | 185 |
| ctarine and peach freestone) | 2.4–3.3 | 1.0 | 1.2 | 0.2 | 0.3 | 20–80 | 100 |
| ach (cling) | 2.6–3.5 | 1.0 | 1.2 | 0.2 | 0.3 | 20–80 | 100 |
| ar | 2.3–2.8 | 0.7 | 1.0 | 0.25 | 0.3 | 20–70 | 80 |
| m (Japanese) | 2.3–2.8 | 1.0 | 1.1 | 0.2 | 0.3 | 30–60 | 80 |
| ne | 2.3–2.8 | 1.0 | 1.3 | 0.2 | 0.3 | 30–80 | 100 |
| lnut | 2.2–3.2 | 0.9 | 1.2 | 0.1 | 0.3 | 26–200 | 300 |

equate levels for all fruit and nut crops: P, 0.1%–0.3%; Cu, over 4 ppm; Mn, ver 20 ppm; Zn, over 15 ppm.

[1]Leaves are July samples from nonfruiting spurs on spur-bearing trees, fully expanded basal ot leaves on peaches and olives, and terminal leaflets on walnut. Adapted from information of Uriu, J. Beutel and O. Lilleland.

[2]Percentage of N in August and September samples can be 0.2%–0.3% lower than July samples d still be equivalent. Nitrogen levels higher than underlined values will adversely affect fruit ality and tree growth. Maximum N for Blenheims should be 3.0% and for Tiltons, 3.5%.

[3]Potassium levels between deficient and adequate are considered "low" and may cause reduced it sizes in some years. Potassium fertilizer applications are recommended for deficient orchards t test applications only for "low" K orchards.

[4]Excess Na or Cl causes reduced growth at the levels shown. Leaf burn may or may not occur en levels are higher. Salinity problems can be confirmed with soil or root analysis.

## Table 8-8
### Leaf Analysis Guide for Diagnosing Nutrient Status of Mature Valencia and Navel Orange Trees[1]

| Element | | Ranges[2] | | |
| --- | --- | --- | --- | --- |
| | | Deficient Below | Optimum | Excess Above |
| N | % | 2.2 | 2.4 to 2.6 | 2.8 |
| P | % | 0.09 | 0.12 to 0.16 | 0.3 |
| K[3] | % | 0.40 | 0.70 to 1.09 | 2.3? |
| Ca | % | 1.6? | 3.0 to 5.5 | 7.0? |
| Mg | % | 0.16 | 0.26 to 0.6 | 1.2? |
| S | % | 0.14 | 0.2 to 0.3 | 0.6 |
| B | ppm | 21 | 31 to 100 | 260 |
| Fe[4] | ppm | 36 | 60 to 120 | 250? |
| Mn[4] | ppm | 16 | 25 to 200 | 1,000 |
| Zn[4] | ppm | 16 | 25 to 100 | 300 |
| Cu | ppm | 3.6 | 5 to 16 | 22? |
| Mo | ppm | 0.06 | 0.10 to 3.0 | 100? |
| Cl | % | ? | <0.3 | 0.7 |
| Na | % | | <0.16 | 0.25 |
| Li | ppm | | <3 | 35? |
| As | ppm | | <1 | 5 |
| F | ppm | | <1 to 20 | 100 |

[1]With the exception of N values, this guide can be applied to grapefruit, lemon and probably other commercial citrus varieties. Data of Embleton, Jones and Platt in *Soil and Plant-Tissue Testing in California*, 1978.

[2]Based on concentration of elements in 5- to 7-month-old, spring-cycle leaves from nonfruiting terminals. Leaves selected for analysis should be free of obvious tipburn, insect or disease injury, mechanical damage, etc., and be from trees that are not visibly affected by disease or other injury.

[3]Potassium ranges are for effects on number of fruits per tree.

[4]Leaves that have been sprayed with Fe, Mn or Zn materials may analyze high in these elements, but those of the next growth cycle may have values in the deficient range.

# SUPPLEMENTARY READING

1. *Critical Nutrient Ranges in Northwest Crops.* A. J. Dow. Western Reg. Pub. 43. Washington State University. 1980.

2. *Diagnostic Criteria for Plants and Soils.* H. D. Chapman, ed. University of California, Division of Agricultural Sciences. 1966.

3. *Handbook of Reference Methods for Plant Analysis.* J. B. Jones. Soil and Plant Analysis Council. 1992.

4. *Handbook on Reference Methods for Soil Analysis.* J. B. Jones. Soil and Plant Analysis Council. 1992

5. *Hunger Signs in Crops,* Third Edition. H. B. Sprague, ed. David McKay Co., Inc. 1964.

6. "Leaf Analysis as a Diagnostic Tool and Guide to Fertilization." T. W. Embleton, W. W. Jones, C. K. Labanauskas and W. Reuther. *The Citrus Industry,* Vol. III (Revised). University of California, Division of Agricultural Sciences. 1973.

7. *Methods of Soil Analysis,* Part II. A. L. Page, ed. American Society of Agronomy. 1982.

8. *Soil Testing and Plant Analysis.* R. L. Westermann, ed. Soil Sci. Soc. of America. 1990.

9. *Soil and Plant Analysis Laboratory Registry for the U.S. and Canada.* Soil Testing and Plant Analysis Council. 1992.

10. *Soil Testing Procedures for California.* CFA-SIC Publication. 1980.

11. *Soil Testing: Sampling Correlation, Calibration, and Interpretation.* J. R. Braun, ed. Soil Sci. Soc. of America. 1987.

# Chapter 9

# *Correcting Soil Problems with Amendments*

The soils under consideration in this chapter have problems that require special management and remedial measures. The primary concern is with the use of amendments such as lime, gypsum, sulfur, and other materials which, when properly used, make the soil more productive.

Those soils which have excess acidity may produce toxicity in growing crops from solubilized aluminum and manganese. Excessive acidity also alters the populations and activities of microorganisms involved in nitrogen and sulfur transformations in soil, and thus affects the availability of these nutrients to higher plants. Phosphorus availability is greatly reduced in acid soils. Table 9-1 shows acidity tolerance of many different commercial plants. Usually soil acidity is an indication of a low level of calcium and magnesium. The use of lime or other amendments can ameliorate many or all of these problems.

Those soils which have a high pH, often referred to as alkaline soils, may be afflicted with high sodium content, excess salts, poor structure, and other problems. The treatment of these soils with amendments requires a completely different approach. This chapter will be concerned with the identification of the problems and the appropriate corrective measures.

# *ACID SOILS*

In the West, acid soils are generally found in areas of high rainfall, on sandy soils, where high rates of acid-forming fertilizers have been used on poorly buffered soils, or where peat or organic deposits occur. Acid soils are more readily formed from weathering of acid igneous rocks, such as granites, and secon-

Table 9–1
Plants Grouped According to Their Tolerance to Acidity

| Very Sensitive to Acidity | Will Tolerate Slight Acidity | Will Tolerate Moderate Acidity | Strong Acidity Favorable |
|---|---|---|---|
| Alfalfa | Soybean | Vetch | Blueberry |
| Sweet clover | Red clover | Oats | Cranberry |
| Barley | Mammoth clover | Rye | Holly |
| Sugar beet | Alsike clover | Buckwheat | Rhododendron |
| Cabbage | White clover | Millet | Azalea |
| Cauliflower | Timothy | Sudan grass | |
| Lettuce | Kentucky bluegrass | Redtop | |
| Onion | Corn | Bentgrass | |
| Spinach | Wheat | Tobacco | |
| Asparagus | Pea | Potato | |
| Beet | Carrot | Field bean | |
| Parsnip | Cucumber | Parsley | |
| Celery | Brussels sprouts | Sweet potato | |
| Muskmelon | Kale | Cotton | |
| | Kohlrabi | Peanuts | |
| | Pumpkin | | |
| | Radish | | |
| | Squash | | |
| | Lima, pole and snap beans | | |
| | Sweet corn | | |
| | Tomato | | |
| | Turnip | | |
| | Sorghum | | |

dary rocks, such as sandstones, than from basic igneous rocks, such as basalts.

Soils become acid because the cations on the soil colloids, primarily calcium, are replaced by hydrogen ions. This implicates ion exchange and other adsorptive reactions that are associated with the colloids, such as the type of clay, the amount of organic matter, etc. Aluminum also comes into play to the extent that highly acid soils can result from aluminum saturation.

The nature of soil acidity is complex; one part, called the active acidity, is made up of the hydrogen ions in the soil solution. These are the ions measured when the pH of a soil is determined. Another and much larger part of the total acidity is usually referred to as the potential acidity, representing the hydrogen ions held on colloidal surfaces. Since clays and organic particles have large surface areas, they usually have a much higher total acidity than sandy soils, and it takes more amendment to change the pH of these soils.

The pH measurement is commonly made when testing soils. Although it is a useful index, it is often misunderstood and misused. The measurement is usually made on a 1:1 mixture of soil and water or on a saturated soil paste. Differences of several tenths of a pH unit are observed as the water:soil ratio is changed. The interpretation of soil acidity must be based upon more than a simple pH reading.

The pH range for acid soils ordinarily is from 4.0 to 7.0. Values much below 4.0 are obtained only when free acids, such as sulfuric acid, are present. Values above 7.0 indicate alkalinity.


## LIME REQUIREMENT

Different methods have been developed to determine the amount of lime needed to bring the pH of an acid soil to a desirable range. All of those presently used take into consideration the buffering capacity of the soil. Table 9-2 shows what

**Table 9–2**
**Amount of Limestone Needed to Change**
**the Soil Reaction (Approximate)[1]**

| Change in pH Desired in Plow- Depth Layer | Pounds of Limestone (CaCO₃) per Acre[2] | | | | | |
|---|---|---|---|---|---|---|
| | Sand | Sandy Loam | Loam | Silt Loam | Clay Loam | Muck |
| 4.0 to 6.5 | 2,600 | 5,000 | 7,000 | 8,400 | 10,000 | 19,000 |
| 4.5 to 6.5 | 2,200 | 4,200 | 5,800 | 7,000 | 8,400 | 16,200 |
| 5.0 to 6.5 | 1,800 | 3,400 | 4,600 | 5,600 | 6,600 | 12,600 |
| 5.5 to 6.5 | 1,200 | 2,600 | 3,400 | 4,000 | 4,600 | 8,600 |
| 6.0 to 6.5 | 600 | 1,400 | 1,800 | 2,200 | 2,400 | 4,400 |

[1]A dolomitic limestone which contains both Ca and Mg is preferable wherever there is a possible lack of magnesium.

[2]Assumes a complete reaction.

the approximate effect of finely ground limestone is on different soils.

## LIMING MATERIALS

The materials commonly used for liming soils are the carbonates, oxides, hydroxides and silicates of calcium and magnesium. A material does not qualify as a liming compound just because it contains calcium or magnesium. *Gypsum, for example, has little direct effect on soil pH and cannot be used to correct a low soil pH.* Table 9-3 gives the liming materials most commonly used to treat acid soils.

The value of a liming material is governed by molecular composition, purity and degree of fineness. Some established standards indicate that all material must pass through a 60-mesh screen to have a full efficiency rating. Smaller particle sized liming materials (minus 100 mesh) will react more rapidly, but generally are more difficult to apply. The use of ash from wood-fueled power plants has been shown by University of California Extension personnel to be effective in correcting soil acidity.

Table 9–3
Common Liming Materials[1]

| Name | Chemical Formula | Equivalent % CaCO$_3$ | Source |
|---|---|---|---|
| Shell meal | CaCO$_3$ | 95 | Natural shell deposits |
| Limestone | CaCO$_3$ | 100 | Pure form, finely ground |
| Hydrated lime | Ca(OH)$_2$ | 120–135 | Steam burned |
| Burned lime | CaO | 150–175 | Kiln burned |
| Dolomite | CaCO$_3$•MgCO$_3$ | 110 | Natural deposit |
| Sugar beet lime[2] | CaCO$_3$ | 80–90 | Sugar beet by-product lime |
| Calcium silicate | CaSiO$_3$ | 60–80 | Slag |
| Power plant ash | CaO, K$_2$O | 25–50 | Wood-fired power plants |
| Cement kiln dust | CaO, CaSiO$_3$ | 40–60 | Cement plants |

[1]Ground to same degree of fineness and at similar moisture content.
[2]From 4% to 10% organic matter.

The benefits of liming acid soils are broad in scope. Calcium and magnesium are essential plant nutrients, and their addition may provide direct value. Additionally, the correction of adverse chemical, physical, and biological conditions may result in striking improvements in plant growth. Phosphorus availability is greatest at a soil pH of 6.5 to 7.0. Toxicities of elements are reduced at pH values of over 6.0, and availability of the micronutrients is optimized at pH 5.5-6.5. Biological activity is favored at a neutral or near neutral pH, including such processes as nitrification, nitrogen fixation, decomposition of plant residues, etc. At the same time, soil aggregation and good structural development are favored.

# SALINE AND SODIC (ALKALI) SOILS

Saline soils generally occur in arid or semiarid regions. In humid areas, rainfall is usually sufficient to move the soluble salts out of the soil, making the incidence of salinity rare. However, the intrusion of sea or brackish water may induce soil salinization, especially in low-lying or coastal areas. Arid regions frequently are inadequately drained and are subject to high evaporation rates, thus allowing salt buildup to occur. Often irrigation is practiced where there are no drainage outlets, and this can result in saline or sodic problems.

The soluble salts that occur come indirectly from the weathering of primary minerals and from water which carries salts from other locations. For example, the Colorado River at Yuma, Arizona, carries more than 1 ton of salt per acre-foot of water. Using this water will result in rapid buildup of salt unless adequate drainage is provided and proper irrigation practices are used.

Sodic (alkali) soils contain excessive amounts of sodium. Irrigating with water containing a high proportion of sodium

Figure 9-1. Salt buildup in a California vineyard.

Figure 9-2. Even a salt-tolerant crop such as barley may be affected by salt accumulation.

will result in sodium replacing calcium and magnesium on the clay. As a consequence of this adsorption of sodium, sodic (alkali) soils are formed.

## SALINE SOILS

This term is applied to soils which have a conductivity of the saturation extract ($EC_e$) greater than 4.0 dS/m, and an exchangeable sodium percentage (ESP) less than 15 percent. These soils normally have a pH value below 8.5 and have good physical properties.

Reclamation of these soils can be accomplished by leaching with high quality irrigation water. Chemical amendments are usually not required. Successful reclamation requires adequate drainage and using the appropriate amount of water. See Chapter 2 for leaching guidelines.

**Figure 9-3. Spreading finely ground sulfur as a soil amendment.**

## SODIC (ALKALI) SOILS

This term is applied to soils which have an $EC_e$ less than 4.0 dS/m, and an ESP greater than 15 percent. They normally have a pH greater than 8.5 and are characterized by their particularly poor physical structure. Sodic soils contain sufficient exchangeable sodium to interfere with the growth of most crops. These soils are commonly termed *alkali, black alkali* and *slick spot* soils. The darkened appearance is caused by the dispersed and dissolved organic matter deposited on the soil surface by evaporation.

Adsorbed sodium causes disintegration of the soil aggregates, dispersing the soil particles and effectively reducing the large pore space. This makes leaching difficult since the soil becomes almost impervious to water.

On non-calcareous soils, gypsum or other soluble calcium salts must be applied. Another approach sometimes used is to apply elemental S with a liming material. On calcareous soils, treatment may be with gypsum or acidifying materials, which in turn solubilizes native calcium in the soil. The soluble calcium replaces sodium on the clay surface and helps bring about a better physical condition that will allow sodium and excess salts to be leached. Organic materials such as manure, crop residues, etc., may be helpful by providing a better physical condition for leaching (more porous soil).

Figure 9-4. Using a gypsum pit for treating high-sodium water.

## SALINE-SODIC SOILS

This term refers to soils with an $EC_e$ above 4.0 dS/m and an ESP greater than 15 percent. Such soils normally have fair to poor physical properties. Water penetration is normally slower than on saline soils. Otherwise, these soils exhibit characteristics of both saline and sodic soils.

Saline-sodic soils sometimes contain gypsum. If this is the

case, leaching will help dissolve calcium, and the soil will provide its own gypsum amendment. Free calcium carbonate (limestone) may also occur in these soils. The selection of amendments in this case can also be made from those that solubilize calcium carbonate and form soluble calcium in the soil. Such amendments are elemental sulfur, sulfuric acid, ferric and ferrous sulfate, aluminum sulfate, etc. Good drainage and leaching are required to remove the sodium as discussed in previous sections.

# SOIL AMENDMENTS

Identification of the specific problem is essential before amendments are chosen. Efforts are often made to amend a saline or sodic soil without awareness that the condition may be accompanied by other chemical or physical problems in the soil. Also, there are those who claim that some special material or process will magically cure the problem. The principles concerning the use and selection of soil amendments are well known, and no shortcut to the proper application of these principles will bring any effective or lasting benefits.

## SOIL CONSIDERATIONS

The presence of free lime (calcium carbonate) in the soil allows the widest selection of amendments. To test for this, a simple procedure can be followed by taking a spoonful or clod of soil and dropping a few drops of muriatic or sulfuric acid on it. If bubbling or fizzing occurs, this indicates the presence of carbonates or bicarbonates. A quantitative determination of lime content requires a laboratory analysis.

If the soil contains lime, any of the amendments listed in Table 9-4 may be used. If lime is absent, select only those amendments containing soluble calcium.

Table 9-4
Commonly Used Materials and Their
Equivalent Amendment Values

| Material (100% Basis) | Chemical Formula | Tons of Amendment Equivalent to | |
|---|---|---|---|
| | | 1 Ton of Pure Gypsum | 1 Ton of Soil Sulfur |
| Gypsum | $CaSO_4 \cdot 2H_2O$ | 1.00 | 5.38 |
| Soil sulfur | S | 0.19 | 1.00 |
| Sulfuric acid (conc.) | $H_2SO_4$ | 0.61 | 3.20 |
| Ferric sulfate | $Fe_2(SO_4)_3 \cdot 9H_2O$ | 1.09 | 5.85 |
| Lime sulfur (22% S) | $CaS_x$ | 0.68 | 3.65 |
| Aluminum sulfate | $Al_2(SO_4)_3 \cdot 18H_2O$ | 1.29 | 6.94 |
| Ammonium polysulfide | $(NH_4)_2S_x$ | 0.37 | 1.95 |

The percent purity is given on the bag or identification tag.

## TYPES OF AMENDMENTS

Soluble calcium amendments, such as gypsum, react in the soil as follows:

gypsum + sodic soil → calcium soil + sodium sulfate

Leaching is essential to remove the sodium salt.
Acids, such as sulfuric acid, require two steps:

1. sulfuric acid + lime → gypsum + carbon dioxide + water

2. gypsum + sodic soil → calcium soil + sodium sulfate

The acid-forming materials, such as sulfur, go through three steps. The first step is oxidation to form the acid, then steps 2 and 3 are the same as 1 and 2 above.

1. sulfur + oxygen + water → sulfuric acid

2. sulfuric acid + lime → gypsum + carbon dioxide + water

3. gypsum + sodic soil → calcium soil + sodium sulfate

For information on using elemental S, see Table 9-5.

**Table 9–5**
**The Approximate Amounts of Elemental Sulfur (99%) Needed
to Increase the Acidity of the Plow-Depth Layer
of a Carbonate-Free Soil**

| Change in pH Desired | Pounds of Sulfur per Acre | | |
|:---:|:---:|:---:|:---:|
| | Sand | Loam | Clay |
| 8.5 to 6.5 | 2,000 | 2,500 | 3,000 |
| 8.0 to 6.5 | 1,200 | 1,500 | 2,000 |
| 7.5 to 6.5 | 500 | 800 | 1,000 |
| 7.0 to 6.5 | 100 | 150 | 300 |

Note: Substantially larger amounts of sulfur are required for soils containing free calcium carbonate.

## *EFFECTIVENESS OF AMENDMENTS*

The values given in Table 9-4 are for 100 percent pure amendments. If an amendment is not pure, a simple calculation will indicate the amount needed to be equivalent to 1 ton of pure material:

$$\frac{100}{\% \text{ purity}} = \text{tons of material}$$

Example: If gypsum is 60 percent pure, the calculation would be $100/60 = 1.67$ tons, or 1.67 tons of 60 percent gypsum would be equivalent to 1.00 ton of 100 percent pure gypsum.

When considering sulfur, the purity and degree of fineness must be taken into account. Most sulfur is over 99 percent pure. Elemental sulfur must be incorporated and oxidized by microorganisms before it is effective as an amendment. The finer the material or greater the surface area, the faster it will be oxidized in the soil. Finely divided sulfur particles may oxidize within one season, while coarse materials may take years.

# MANAGEMENT OF
# SALINE AND SODIC SOILS

It may not be practical to completely reclaim saline or sodic soils or even to maintain these soils at a low saline or sodic condition. The reasons may be cost of reclamation, inability to adequately drain, high cost of amendments, low quality irrigation water, etc.

Practices that aid in the management of salt and sodium include:

1. Selection of crops or crop varieties that have tolerance to salt or sodium.
2. Use of special planting procedures that minimize salt accumulation around the seed.
3. Use of sloping beds or special land preparation procedures and tillage methods that provide a low salt environment for the germinating seed.
4. Use of irrigation water to maintain a high water content to dilute the salts or to leach the salts away from the germination and root growing zone.
5. Use of physical amendments for improving soil structure.
6. Deep ripping the soil to break up hardpan or other impervious layers to provide internal drainage.
7. Use of chemical amendments as described.
8. Establishment of proper surface and internal drainage.

It is essential to know the nature of the soil, both physical and chemical, the quality and quantity of irrigation water available, the climate of the area including the growing season, the

**Figure 9-5. Wise usage of soil amendments and fertilizers produces high-yielding crops. (Onion harvest in the Imperial Valley, California.)**

economics of the situation, etc., before a satisfactory management program can be developed. Consulting with qualified soil scientists and having appropriate tests conducted are essential steps in reclamation and management.

## SUPPLEMENTARY READING

1. *Diagnosing Soil Salinity.* USDA Agricultural Information Bulletin 279. 1963.

2. *Diagnosis and Improvement of Saline and Alkali Soils.* USDA Agricultural Handbook No. 60. 1954.

3. *Gypsum and Other Chemical Amendments for Soil Improvement.* University of California, Extension Leaflet 149. 1962.

4. *Irrigation of Agricultural Crops.* Agronomy Monograph No. 30. American Society of Agronomy. 1990.

5. *Reclaiming Saline and Alkali Soils.* Extension Publication, Fresno Co., CA. 1972.

6. *Recycling Wood Fueled Co-generation Fly Ash on Agricultural Lands.* Meyer, Roland D. et al. University of California Cooperative Extension. 1990.

7. *Soil Acidity and Liming.* Agronomy Monograph 12. American Society of Agronomy. 1991.

8. *Soil Survey Manual.* USDA Agricultural Handbook No. 18. Soil Survey Staff. U.S. Government Printing Office. 1951.

9. *Soil, Yearbook of Agriculture.* U.S. Government Printing Office. 1957.

# Chapter 10

# *Best Management Practices*

The development of agricultural technology has allowed producers to turn to improved systems to achieve production objectives with environmentally responsible methods. These production systems utilize Best Management Practices (BMPs). These BMPs combine scientific research with practical knowledge to optimize yields and increase crop quality, utilizing increased fertilizer efficiency and minimizing input loss. The adoption of BMPs provides the greatest economic return for the producer, while maintaining environmental soundness of production methods.

A large number of BMPs have been developed by academic, governmental, and private industry researchers to reflect practices in all phases of agricultural production. The best managers incorporate BMPs for all phases of production. Maintaining or increasing production levels must be the end result of an agricultural system that relies on fewer acres and increasing demand, yet integrity of the environment must remain intact. The adoption of BMPs will lead to these goals.

Adequate food and fiber supplies are a basic need for mankind. Productive agriculture provides these supplies and plays an important role in the economic growth of developing countries. Yet it also has an impact on the environment. Recognizing that there will **never** be zero risk in food production or any other human activity, agricultural leaders must seek the establishment of practical farming systems which are sustainable in both environmental and economic terms. A fundamental requirement is therefore to recognize that, while the major principles of crop nutrition apply worldwide, they must be carefully interpreted to meet the specific climatic, soil, and management conditions experienced by the individual farmer. BMPs are site-specific, utilizing proven technology that is modified and amended to fit individual soil types within a particular cropping area. They must be flexible guidelines - not strict rules - that might be limited to a single farm or could be extended throughout an entire country. It seems obvious that no simple, single solution to production and/or environmental situations can be regulated or legislated, rather, the need is for adoption of site-specific BMPs within modern production agriculture. Factors associated with BMPs include:

- BMP technology should be based on practical field research.

- BMPs should be continually improved through implementation of current research.

- BMPs lead to maximum economic yields.

- BMPs include soil and water conservation and sound agronomic practices.

- BMPs encourage high input efficiency.

- BMPs are flexible to accommodate specific farm, soil, and crop conditions.

- BMP adoption depends on strengthening the educational base of agriculture.

- BMPs provide the opportunity to better manage environmentally-sensitive areas.

- BMPs improve food safety and quality.

- BMPs promote good environmental stewardship.

The purpose of this chapter is to provide several examples

of both broad spectrum and site specific BMPs that have been developed for improved soil fertility and plant nutrition. Pest management, variety selection, crop rotation, and water management, to name a few, will also impact environmentally sound agricultural production. The reader is advised to consult the recommended reading list at the end of this chapter for specific BMPs or contact local cooperative extension offices, USDA-Soil Conservation Service field offices, agricultural consultants, and fertilizer dealers for additional information.

## GOAL ORIENTED BMP'S

1. **Soil testing** has always been known as a tool for determining the most economical rate of fertilizer and soil amendment, based on identifiable levels of plant nutrients in the soil and soil chemical characteristics. It is a valuable BMP for environmentally-responsible farming.

Figure 10-1. Soil sampling is a BMP.

Following recommendations based on soil testing assures that nutrients are applied to soils for good reasons and are backed up with good science. Refer to Chapter 8 for correct procedures for soil sampling.

Practices based on soil testing benefit the environment in several important ways:

- Protect water quality by reducing excess nutrients.

- Conserve finite resources by recommending nutrient additions only when necessary.

- Reduce soil erosion by promoting early root growth and canopy cover.

- Enhance natural resistance to disease and pests through improved crop nutrition.

- Maintain sustainability of agricultural lands through selection of proper soil amendments.

2. **Nutrient management** plans are designed to precisely balance nutrient applications with crop fertility needs in order to enhance water quality and increase farm profitability. Nutrient management is a complex process affected by weather conditions, production objectives, equipment availability, and timing of nutrient applications. Crop advisors and other resource persons must work closely with farmers to develop plans tailored for each farm. A nutrient management plan should be a part of every farm's soil conservation and water quality plan.

Benefits to the farmer are:

- Reduced risk of polluting surface and ground water as a result of excess nutrient applications.

- Efficient integration of commercial fertilizers and other nutrient sources such as manure or sewage sludge, reducing fertilizer costs by taking full advantage of other nutrient sources.

- Improved fertilizer recommendations, which help achieve maximum economic yields.

3. **Conservation tillage** BMPs conserve soil and water resources and improve the efficiency of production inputs such as fuel, fertilizer, and labor. The environmental benefits of conservation tillage stem from reduced soil surface disturbance and maintenance of crop residues on the soil surfaces. Rainfall infiltrates soils and does not runoff as easily under conservation tillage systems as under conventional tillage systems. By curtailing runoff, erosion is reduced and contamination of ground water and surface waters with nutrients, sediments, and crop protection agents is greatly reduced. Conservation tillage is a flexible, broad-based BMP that is widely recommended. Site specific BMPs such as the use of starter fertilizers, water management, and timing of nitrogen applications are necessary to ensure a program that is agronomically sound, economically efficient, and environmentally beneficial.

## SITE SPECIFIC BMPs

Site specific BMPs require specific actions of an operator to achieve the goals of more broad-based BMPs. There are numerous site specific BMPs. Three examples follow:

1. Rate of nitrogen fertilizer applied will be the amount necessary to meet projected crop needs. Potentially leachable nitrogen increases rapidly when the amount applied exceeds that required to attain maximum or near maximum yield. Factors that complicate the optimum nitrogen rate determination are the various sources, losses, and transformations of plant available forms. See Chapters 4 and 5 for more information.

2. Application of nitrogen fertilizer should be timed to coincide with the periods of maximum crop uptake. Nitrogen uptake by crops varies during the growing season. The maximum potential uptake rate is determined by the stage of growth and genetic characteristics of a given crop. A second factor to be considered is the possible time lag between the application of nitrogen and its conversion to a plant available form. The most effective management strategy is one that recognizes crop demand for nitrogen and the release characteristics of all nitrogen sources in the system such that adequate but not excessive levels of soil nitrogen are provided throughout the growing season.

3. Irrigation and nutrient application BMPs must be coordinated to minimize nutrient and water losses. Providing adequate irrigation water for the evaporation and transpiration losses must be integrated with other require-

**Figure 10-2. Petiole sampling for nitrogen and other nutrients.**

Figure 10-3. Laser leveling for more uniform water application.

Figure 10-4. Drip irrigating cotton in Arizona.

ments, such as leaching of excess salts. Growers must be able to accurately determine crop water use through plant and soil water measurements or by estimations based on weather data. Any irrigation scheduling technique must recognize crop-specific soil moisture requirements. Application method and frequency should be selected and adjusted for precise water management. See Chapter 2 for a thorough discussion of irrigation practices.

## SUMMARY

One of the keys to successful implementation of Best Management Practices is that they are flexible combinations of site specific guidelines for each soil, crop, and climate. The goal of maintaining environmental integrity while providing the greatest economical return can only be realized when the best possible practices, those that have been proven to work, are adopted.

## SUPPLEMENTARY READING

1. *Best Management Practices* Booklets. Unocal Corporation, Los Angeles.

2. *EPA Proposed Guidance Specifying Management Measures for Sources of Nonpoint Pollution in Coastal Waters.* U.S. Environmental Protection Agency, Office of Water. May 1991.

3. *Fertilizer Management for Today's Tillage Systems.* The Potash & Phosphate Institute. Norcross, Georgia. 1990.

4. *Improving Fertilizer and Chemical Efficiency Through "High Precision Farming."* R. D. Munson and C. F. Runge. Center for International Food and Agricultural Policy. University of Minnesota, St. Paul, Minnesota. Sept. 1990.

5. *Nitrogen Fertilizer Management in Arizona.* T. A. Doerge, R. L. Roth, and B. R. Gardner. The University of Arizona, Tucson, Arizona. May 1991.

6. *Starter Fertilizer and High Residue: A Profit-Building Combination.* The Potash & Phosphate Institute. Norcross, Georgia. Pamphlet #02-1250.

7. *Sustainable Dryland Argroecosystem Management.* G. A. Peterson, D. G. Westfall, L. Sherrod, R. Kolberg, and B. Rouppet. Colorado Agric. Exp. Station, Fort Collins, and USDA-ARS National Resource Research Center, Great Plains Systems Research Unit, Fort Collins, Colorado. 1993.

# Chapter 11

# Fertilizers and
# the Environment

Research consistently demonstrates that good fertilizer practices that match fertilizer inputs to crop requirements will achieve high, economically sustainable yields, and quality. Such applications are also beneficial to the environment. However, improper or excessive application of fertilizer can lead to potential environmental problems.

Fertilizer production, transportation, storage, and field application are closely regulated by a number of federal and state agencies. Further information may be retrieved from local government or industry association sources.

## FERTILIZERS FEED THE WORLD
## AND MAINTAIN THE ENVIRONMENT

In a position paper presented at the United Nations conference on Environment and Development in Rio de Janeiro, 1992, it was stated that:

- by 2025 world population would jump by 50 percent, to 8.5 billion.

- three-quarters of this number would live in developing countries: Asia alone will have to feed an extra 600 million people.

- the land available for growing food will diminish as population growth continues. The area of cultivated land per person is estimated to fall to below 0.32 acres by 2000, down from 0.37 acres in 1992.

Quite simply, more food must be produced from less land. Proven cropping management and input technologies, including fertilizers, will be essential to feeding the world. Importantly, sound nutrient management is entirely consistent with environmental protection and preserving the agricultural resource for future generations. In fact, fertilizers are a critical component of sustainable agriculture.

## *FERTILIZER FACTS*

Fertilizers provide essential plant nutrients to supplement the soil supply which often falls short of crop demand. This relationship has been known for centuries and has grown in importance as growth in world population has increased demands for food and fiber.

Environmental risks associated with fertilizer use stem from poor management of application method, rate, and timing. Nitrogen and phosphorus are the nutrients most commonly associated with environmental concerns. Nitrogen in the form of nitrate can enter water supplies where it poses possible human health risks. Phosphorus contained, for example, in run-off from agricultural lands may stimulate an over-abundance of algal growth in surface waters. As a result, clearly defined, safe levels of nutrients in water supplies are established and monitored in the United States and elsewhere. Although fertilizer derived nutrients may contribute to elevated levels in some instances, many naturally occurring soil derived nutrients will also appear in water samples. For example:

- Recent U.S. Environmental Protection Agency (EPA) surveys indicate that a very small proportion of wells sampled (1.2 percent public, 2.4 percent private) contain nitrate above the maximum contaminant level. Such incidences were considered site specific and not widespread. Furthermore, nitrate contamination from improper manure management, septic tanks, and waste water treatment plants was also implicated.

• Many university studies across the United States clearly show that correct fertilizer nitrogen use cannot be correlated with increases in ground or surface water nitrate levels. In Illinois, water analyses from the mid-1800s indicate nitrate levels well in excess of the public health standard, at a time when agriculture was in its infancy.

• In predominately irrigated agriculture, such as the intermountain west, studies indicate that the practice of irrigating crops may actually remove phosphorus and nitrogen from stream waters.

Plants will not discriminate between nutrients provided by soil processes, organic sources such as animal wastes, or synthetic fertilizers. Nitrate is nitrate, for example. In fact, some research data strongly suggest that there may be more environmental risk associated with manure and legume nitrogen management compared with fertilizers. (Table 11-1)

### Table 11-1
### Nitrate Leaching Potential from Various Nitrogen Sources

| Nitrogen Source | Soil Nitrate Content |
|---|---|
| | *(ppm)* |
| No nitrogen | 8 |
| Commercial nitrogen | 10 |
| Animal manure | 49 |
| Alfalfa plowed down | 62 |

(Michigan State University)

Additional management difficulties such as low concentration, handling, storage and transportation logistics, and inappropriate application timing are associated with animal waste utilization in crop production.

Fertilizers promote soil health. Long-term studies from around the world indicate that fertilizers contribute in a positive manner to the sustainability of agricultural soils. At the world's oldest agricultural research establishment, Rothamsted, England, fields receiving commercial fertilizer for over 150 years are more productive now than at any time in their past. Studies from Australia, New Zealand, Canada, and the United States also show that fertilizer has rendered no long term harm to soil organic components, which were mostly improved through fertilizer use.

Commercial fertilizer use has paralleled increases in crop production. However, fertilizer use efficiency based on applied amounts per unit of production has increased markedly. An analysis of fertilizer usage and corn production over a thirty year period has been conducted by Potash & Phosphate Institute scientists. This study clearly shows that U.S. farmers are now producing more bushels per pound of applied nutrient than ever before, and that the trend continues. This is a remarkable feat since corn yields have nearly doubled, yet fertilizer use is now declining for the most part. Refinements in crop management and emerging technologies suggest that improvements in fertilizer use efficiency will continue.

Fertilizers should not be confused with agricultural chemicals. Pest control chemicals (weeds, diseases, insects, etc.) are predominantly synthetic molecules active in low concentrations and often crop and pest specific. Safely handled and applied, pesticides are essential tools in present day agriculture.

Commercial fertilizers are refined or upgraded, safe compounds of naturally occurring elements. Various fertilizer application methods can actually assist in plant disease tolerance and increase the efficacy of applied agricultural chemicals.

# POSITIVE IMPACTS
# OF PROPER FERTILIZER USE
# ON THE ENVIRONMENT

Fertilizer improves and protects the environment in several ways:

1. Reduces soil erosion to maintain soil productivity and reduce pollution of surface waters.

2. Is key to producing efficient rooting systems to help reduce pollution of ground waters.

3. Greatly improves land use efficiency.

4. Assists in the safe disposal of degradable wastes and land remediation/reclamation.

5. Sustains green top growth, essential to gaseous exchange.

## REDUCES SOIL EROSION

Adequately fertilized crops will have both extensive root systems and above ground growth (tops). A well developed top growth reduces the pounding effect of water drops from rain or sprinklers. The energy of the drops is dissipated so that instead of disrupting the soil surface, the water is allowed to penetrate the soil. As a result, run-off is reduced and erosion is minimized. Similarly, extensive root systems as a result of good fertility will help to hold soils in place and decrease the potential for soil loss in water run-off.

The expansion of minimum/zero tillage (minor or no soil tillage prior to crop planting) is reliant upon fertilizer use to increase levels of crop residue that aid moisture conservation and reduce soil erosion.

Soil erosion is a major factor in phosphate loss from soils. Phosphorus is carried off with soil particles in run-off water as adsorbed and precipitated phosphate. Since soil sediment also

# Well Fertilized Crops Reduce Erosion

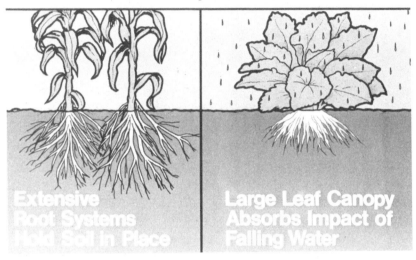

Figure 11-1. Leaf and root systems reduce soil erosion.

Figure 11-2. Soil erosion can be serious when the soil is not held in place by roots of well-fertilized crops.

contains nitrogen and other nutrients, any reduction in erosion will reduce the potential for increased levels of these nutrients in surface waters.

## IMPROVED ROOT SYSTEMS

A diverse research database indicates that properly fertilized crops have extensive rooting systems. Such rooting systems are more nutrient (both soil and applied) and water efficient. Studies from the Great Plains and elsewhere consistently demonstrate the role of fertilizers in sustaining nutrient and water use efficiency of various rotations. (Table 11-2)

### Table 11–2
### Nitrogen Fertilizer Use, Residual Nitrate and
### Water Use Efficiency of Three Crop Rotations[1]

| Rotation[2] | After 4 Years | | Grain/ Precipitation |
| --- | --- | --- | --- |
| | Fertilizer N | Residual Nitrate-N | |
| | *(lb/ac)* | *(lb/ac – 6 ft)* | *(lb/in)* |
| WF | 60 | 134 | 65 |
| WCF | 135 | 100 | 133 |
| WCMF | 112 | 44 | 138 |

[1]Westfall & Peterson, Colorado State University, 1990.
[2]W, C, M, F = wheat, corn, millet, fallow, respectively.

## IMPROVED LAND USE EFFICIENCY

An ever-increasing population continues to put pressure on land space through urbanization and the need for recreational areas. More food and fiber will have to be produced from a diminishing agricultural land area. For example, urbanization is the driving force behind land use conversion in the state of California. Based on current population trends, it has been estimated that another 1-2 million acres of land will be urbanized by 2000.

Figure 11-3. Active root systems intercept nutrients before they reach groundwater.

Figure 11-4. Almond production in northern California. Skilled irrigation and fertilizer management lead to efficient production.

Loss of prime agricultural acreage could be offset by placing marginal land into production. However, this raises environmental questions since these marginal lands are often less fertile and highly erodible.

For agriculture to continue to feed sustained population growth and accommodate increased land demands, fertilizer management will be an essential tool toward maintaining crop productivity on a diminishing acreage.

## NON-AGRICULTURAL ENVIRONMENTAL BENEFITS OF FERTILIZERS

Sound nutrient management will be essential as agricultural land, together with domestic and landscaped areas are used increasingly to dispose of degradable wastes. These materials include sewage sludge and green manure, such as grass clippings, composted to reduce flows to landfills. Substantial acres will be required to dispose of these wastes. The possibility of widespread agricultural use of municipal wastes as nutrient sources has raised concerns including salt load, heavy metals, and human health risk due to exposure to disease organisms. Restricted plant growth and ground water pollution are additional concerns.

Common commercial fertilizer materials are also used to positive environmental benefit in non-agricultural situations. Some examples include: reclamation of mine sites, bio-remediation of oil spills, fire retardants and fire-fighting materials, and ameliorating heavy metal contamination of soils.

## GASEOUS EXCHANGE

Plant nutrition is essential to maintaining efficient above-ground canopies. Through the process of photosynthesis, green plants utilize atmospheric carbon dioxide and generate life-sustaining oxygen.

# MATCHING FERTILIZER INPUTS
# TO CROP NEEDS

Sound fertilizer usage is essentially providing crops with the right amount, at the right time, and in the right place. Vital management tools include soil, tissue, and water testing; the nutrient requirements of crops and yield histories and expectations for a specific field. Application method and timing of fertilizers will exert a profound effect on the efficiency of any nutrient management program. The reader is referred to Chapters 7 and 10 for a complete review of the multi-faceted nature of integrated crop management, the role of fertilizers, and environmental safety.

# SUPPLEMENTARY READING

1. *Better Crops With Plant Food.* Potash & Phosphate Institute, Atlanta, GA. Fall, 1990.

2. *Facts From Our Environment.* Potash & Phosphate Institute, Atlanta, GA. January, 1991.

3. *Farming, Fertilizers and the Nitrate Problem.* Addiscott, Whitmore, Powlson. C.A.B. International. Rothamsted Experimental Station. 1991.

4. *Fertilizer Feeds the World.* Brochure. The Fertilizer Institute, Washington, D.C.

5. *Managing Nitrogen for Groundwater Quality and Farm Profitability.* Follett, Keeney, and Cruse. Soil Science Society of America. 1991.

6. *Pacific Northwest Conservation Tillage Handbook.* Veseth, Wysocki. STEEP Extension Publication. Idaho, Oregon, Washington.

7. *Ten Facts - Affecting Nitrogen Use for Sustainable Agriculture and Nitrates in Groundwater.* Potash & Phosphate Institute, Atlanta, GA. March 1990.

# Appendix A

# *Model Law Relating to Fertilizer Materials*

## *OFFICIALLY ADOPTED DOCUMENTS*

Note – Although these documents have not been passed into law in all states, the subject matter covered herein does represent the official policy of this Association. NOTE: Tentative actions are in *BOLD ITALICS*, new wording is enclosed in brackets [ ], and deleted material is denoted with strike throughs, ~~ababababab~~.

## UNIFORM STATE FERTILIZER BILL
## (Official 1982)

AN ACT to regulate the sale and distribution of fertilizers in the State of _____ . BE IT ENACTED by the legislature of the State of _____ .

Section 1.    Title

This Act shall be known as the "_____ Fertilizer Law of 19____".

---

Source: Association of American Plant Food Control Officials. Official Publication No. 46. 1993.

Section 2.   Enforcing Official

This Act shall be administered by the _____ of the State of _____ , hereinafter referred to as the "_____".

Section 3.   Definitions of Word and Terms

When used in this Act:

(a)   The term "fertilizer" means any substance containing one or more recognized plant nutrient(s) which is used for its plant nutrient content and which is designed for use or claimed to have value in promoting plant growth, except unmanipulated animal and vegetable manures, marl, lime, limestone, wood ashes, and other products exempted by regulation by the _____ .

    (1)   A "fertilizer material" is a fertilizer which either:

        A.   Contains important quantities of no more than one of the primary plant nutrients: nitrogen (N), phosphorus (P), and potassium (K), or

        B.   Has 85 percent or more of its plant nutrient content present in the form of a single chemical compound, or

        C.   Is derived from a plant or animal residue or by-product or natural material deposit which has been processed in such a way that its content of plant nutrients has not been materially changed except by purification and concentration.

    (2)   A "mixed fertilizer" is a fertilizer containing any combination or mixture of fertilizer materials.

    (3)   A "specialty fertilizer" is a fertilizer distributed for non-farm use.

(4)   A "bulk fertilizer" is a fertilizer distributed in a non-packaged form.

(b)   The term "brand" means a term, design, or trademark used in connection with one or several grades of fertilizer.

(c)   Guaranteed Analysis:

*(1)   Until the _____ prescribes the alternative form of "Guaranteed Analysis" in accordance with the provisions of subparagraph (2) hereof, the term "Guaranteed Analysis" shall mean the minimum percentage of plant nutrients claimed in the following order and form:*

*A.   Total Nitrogen (N)                       _____ %*

*Available ~~Phosphoric Acid~~ [Phosphate] (P₂O₅)          _____ %*

*Soluble Potash (K₂O)                  _____ %*

*(Tentative 1992.)*

*B.   For unacidulated mineral phosphatic material and basic slag, bone, tankage, and other organic phosphatic materials, the total ~~phosphoric acid~~ [phosphate] and/or degree of fineness may also be guaranteed. (Tentative 1992.)*

C.   Guarantees for plant nutrients other than nitrogen, phosphorus, and potassium may be permitted or required by regulation by the _____ . The guarantees for such other nutrients shall be expressed in the form of the element. The source (oxides, salts, chelates, etc.) of such other nutrients may be required to be stated on the application for registration and may be included on the label. Other beneficial substances or compounds, determinable by laboratory methods, also may be guaranteed by permission of the _____ and with the advice of the Director of the Agricultural Experiment Station. When any plant nutrients or other substances or compounds are guaranteed, they shall be subject to inspection and analysis in accord with the methods and regulations prescribed by the _____ .

*(2)* *When the _____ finds, after public hearing following due notice, that the requirement for expressing the guaranteed analysis of phosphorus and potassium in elemental form would not impose an economic hardship on distributors and users of fertilizer by reason of conflicting labeling requirements among the states, he may require by regulation thereafter that the "Guaranteed Analysis" shall be in the following form:*

*Total Nitrogen (N)* _____ %

*Available Phosphorus (P)* _____ %

*Soluble Potassium (K)* _____ %

*Provided, however, that the effective date of said regulation shall be not less than six months following the issuance thereof, and Provided, further, That for a period of two years following the effective date of said regulation the equivalent of phosphorus and potassium may also be shown in the form of ~~phosphoric acid~~ [phosphate] and potash; Provided, however, That after the effective date of a regulation issued under the provisions of this section, requiring that phosphorus and potassium be shown in the elemental form, the guaranteed analysis for ni-*

*trogen, phosphorus, and potassium shall consti-tute the grade. (Tentative 1992.)*

*(d)*  *The term "grade" means the percentage of total nitrogen, available phosphorus or ~~phosphoric acid~~ [phosphate], and soluble potassium or potash stated in whole numbers in the same terms, order, and percentages as in the guaranteed analysis. Provided, however, that specialty fertilizers may be guaranteed in fractional units of less than one percent of total nitrogen, available phosphorus or ~~phosphoric acid~~ [phosphate], and soluble potassium or potash: Pro-vided, further, That fertilizer materials, bone meal, manures, and similar materials may be guaranteed in fractional units. (Tentative 1992.)*

(e)  The term "official sample" means any sample of fertilizer taken by the _____ or his agent and designated as "official" by the _____ .

(f)  The term "ton" means a net weight of two thousand pounds avoirdupois.

*(g)*  *The term "primary nutrient" includes nitrogen, avail-able ~~phosphoric acid~~ [phosphate] or phosphorus, and soluble potash or potassium. (Tentative 1992.)*

(h)  The term "percent" or "percentage" means the percentage by weight.

(i)  The term "person" includes individual, partnership, association, firm, and corporation.

(j)  The term "distribute" means to import, consign, manufacture, produce, compound, mix, or blend fertilizer, or to offer for sale, sell, barter or otherwise supply fertilizer in this state.

(k)  The term "distributor" means any person who distributes.

(l)     The term "registrant" means the person who registers fertilizer under the provisions of this Act.

(m)    The term "licensee" means the person who receives a license to distribute a fertilizer under the provisions of this Act.

(n)     The term "label" means the display of all written, printed, or graphic matter, upon the immediate container, or a statement accompanying a fertilizer.

(o)     The term "labeling" means all written, printed, or graphic matter, upon or accompanying any fertilizer, or advertisements, brochures, posters, television, and radio announcements used in promoting the sale of such fertilizer.

(p)     The term "investigational allowance" means an allowance for variations inherent in the taking, preparation, and analysis of an official sample of fertilizer.

(q)     The term "deficiency" means the amount of nutrient found by analysis less than that guaranteed which may result from a lack of nutrient ingredients or from lack of uniformity. (Official 1985.)


Section 4.    Option A — Registration

(a)     Each brand and grade of fertilizer shall be registered in the name of that person whose name appears upon the label before being distributed in this state. The application for registration shall be submitted to the _____ on a form furnished by the _____ and shall be accompanied by a fee of $_____ per each grade of each brand, except those fertilizers sold in packages of 10 pounds or less shall be registered at a fee of $_____ per each grade of each brand. Upon approval by the ___ _____ a copy of the registration shall be furnished to

the applicant. All registrations expire on _____
each year. The application shall include the following information:

(1)   The brand and grade;

(2)   The guaranteed analysis;

(3)   The name and address of the registrant;

(4)   Net weight. (Official 1988).

(b)   A distributor shall not be required to register each grade
      of fertilizer formulated according to specifications which are
      furnished by a consumer prior to mixing, but shall be
      required to register his firm in a manner and at a fee as
      prescribed in regulations by the _____ and to
      label such fertilizer as provided in Section 5(b).

Section 4.    Option B — Registration and Licensing

(a)   No person whose name appears upon the label of a fertilizer
      shall distribute that fertilizer, except specialty fertilizers, to
      a non-licensee until a license to distribute has been obtained
      by that person from the _____ upon payment of
      a $_____ fee. All licenses expire on the _____
      day of _____ each year.

(b)   An application for license shall include:

      (1)   The name and address of licensee.

      (2)   The name and address of each distribution point in
            the state. The name and address shown on the license
            shall be shown on all labels, pertinent invoices, and
            storage facilities for fertilizer distributed by the licensee
            in this state.

(c)   The licensee shall inform the _____ in writing of

additional distribution points established during the period of the license.

(d)     No person shall distribute in this state a specialty fertilizer until it is registered with the _____ by the distributor whose name appears on the label. An application for each brand and product name of each grade of specialty fertilizer shall be made on a form furnished by the ____ _____ and shall be accompanied by a fee of $_____ per each grade of each brand, except those fertilizers sold in packages of 10 pounds or less shall be registered at a fee of $_____ per each grade of each brand. Labels for each brand and product name of each grade shall accompany the application. Upon the approval of an application by the _____ , a copy of the registration shall be furnished the applicant. All registrations expire on the _____ day of _____ each year.

(e) An application for registration shall include the following:

(1) The brand and grade;

(2) The guaranteed analysis;

(3) Name and address of the registrant;

(4) Net Weight. (Official 1988.)

Section 4.    Option C — Licensing

(a)     No person whose name appears upon the label of a fertilizer shall distribute that fertilizer to a non-licensee until a license to distribute has been obtained by that person from the _____ upon payment of a $_____ fee. All licenses expire on the day of each year.

(b)     An application for license shall include:

(1)     The name and address of licensee.

(2)     The name and address of each distribution point in the state.

The name and address shown on the license shall be shown on all labels, pertinent invoices, and storage facilities for fertilizers distributed by the licensee in this state.

(c)   The licensee shall inform the _____ in writing of additional distribution points established during the period of the license.

Section 5.   Labels

(a)   Any fertilizer distributed in this state in containers shall have placed on or affixed to the container a label setting forth in clearly legible and conspicuous form the following information:

(1)   Net weight;

(2)   Brand and grade: Provided, That the grade shall not be required when no primary nutrients are claimed;

(3)   Guaranteed analysis;

(4)   Name and address of the registrant/licensee.

In case of bulk shipments, this information in written or printed form shall accompany delivery and be supplied to the purchaser at time of delivery.

(b)   A fertilizer formulated according to specifications which are furnished by/for a consumer prior to mixing shall be labeled to show the net weight, the guaranteed analysis, and the name and address of the distributor or registrant/licensee.

Section 6.   Inspection Fees

(a)   There shall be paid to the _____ for all fertilizers distributed in this state to non-registrants/non-licensees an inspection fee at the rate of _____ cents per ton; Provided,

That sales or exchanges between importers, manufacturers, distributors, or registrants/licensees are hereby exempted.

(b)     Every registrant/licensee who distributes fertilizer in the state shall file with the _____ a (monthly, quarterly, or semi-annual) statement for the reporting period setting forth the number of net tons of each fertilizer so distributed in this state during such period. The report shall be due on or before thirty days following the close of the filing period and upon such statement shall pay the inspection fee at the rate stated in paragraph (a) of this section. If the tonnage report is not filed and the payment of inspection fees is not made within 30 days after the end of the specified filing period, a collection fee, amounting to 10 percent (minimum $10) of the amount due, shall be assessed against the registrant/licensee and added to the amount due.

(c)     When more than one person is involved in the distribution of a fertilizer, the last person who has the fertilizer registered (is licensed) and who distributed to a non-registrant/licensee dealer or consumer is responsible for reporting the tonnage and paying the inspection fee, unless the report and payment is made by a prior distributor of the fertilizer.

(d)     On individual packages of fertilizer containing 10 pounds or less there shall be paid, in lieu of the inspection fee of _____ cents per ton and in lieu of $_____ per brand and grade, an annual registration and inspection fee of $_____ for each grade of each brand sold or distributed. Where a person distributes fertilizer in packages of 10 pounds or less and in packages over 10 pounds, the annual fee shall apply only to that portion distributed in packages of 10 pounds or less.

(e)     Fees so collected shall be used for the payment of the

costs of inspection, sampling and analysis, and other expenses
necessary for the administration of this Act.

## Section 7.   Tonnage Reports

(a)   The person distributing or selling fertilizer to a non- regis-
trant/non-licensee shall furnish the _____ a report
showing the county of the consignee, the amounts (tons)
of each grade of fertilizer, and the form in which the
fertilizer was distributed (bags, bulk, liquid, etc.). This information shall be reported by one of the following methods:

    (1)   Submitting a summary report approved by the ____
        _____ on or before the _____ day of
        each month covering shipments made during the pre-
        ceding month; or

    (2)   Submitting a copy of the invoice within _____
        business days after shipment.

(b)   No information furnished the _____ under this
section shall be disclosed in such a way as to divulge the
operation of any person.

## Section 8.   Inspection, Sampling, Analysis

(a)   It shall be the duty of the _____ , who may act
through his authorized agent, to sample, inspect, make
analyses of, and test fertilizers distributed within this state
and inspect the storage of bulk fertilizer at any time and
place and to such an extent he may deem necessary to
determine whether such fertilizers are in compliance with
the provisions of this Act. The _____ , individually
or through his agent, is authorized to enter upon any public
or private premises or carriers during regular business hours
in order to have access to fertilizer subject to provisions
of this Act and the regulations pertaining thereto, and to

the records relating to their distribution and storage. (Official 1990.)

(b)   The methods of sampling and analysis shall be those adopted by the Association of Official Analytical Chemists (AOAC). In cases not covered by such methods, or in cases where methods are available in which improved applicability has been demonstrated, the _____ may adopt such appropriate methods from other sources.

(c)   The _____ , in determining for administrative purposes whether any fertilizer is deficient in plant food, shall be guided solely by the official sample as defined in paragraph (e) of Section 3, and obtained and analyzed as provided for in paragraph (b) of this section.

(d)   The results of official analysis of fertilizers and portions of official samples shall be distributed by the _____ as provided by regulation. Official samples establishing a penalty for nutrient deficiency shall be retained for a minimum of 90 days from issuance of a deficiency report.

Section 9.   Plant Food Deficiency

*(a)   Penalty for nitrogen, available ~~phosphoric acid~~ [phosphate] or phosphorus, and soluble potash or potassium — If the analysis shall show that a fertilizer is deficient (1) in one or more of its guaranteed primary plant nutrients beyond the investigational allowances and compensations as established by regulation, or (2) if the overall index value of the fertilizer is below the level established by regulation, a penalty payment of _____ times the value of such deficiency or deficiencies shall be assessed. When a fertilizer is subject to a penalty payment under both (1) and (2), the larger penalty payment shall apply. (Tentative 1992.)*

(b)     Penalty payment for other deficiencies — Deficiencies beyond
        the investigational allowances as established by regulation
        in any other constituent(s) covered under Section 3 paragraph
        (c)(1) B and C of this Act, which the registrant/licensee is
        required to or may guarantee, shall be evaluated and penalty
        payments prescribed by the _____ .

(c)     All penalty payments assessed under this section shall be
        paid by the registrant/licensee to the consumer of the lot
        of fertilizer represented by the sample analyzed within three
        months after the date of notice from the _____
        to the registrant/licensee, receipts taken therefore and
        promptly forwarded to the _____ . If said consumer
        cannot be found, the amount of the penalty payments shall
        be paid to the _____ who shall deposit the same
        in the appropriate state fund allocated to fertilizer control
        service. If, upon satisfactory evidence, a person is shown
        to have altered the content of a fertilizer shipped to him
        by a registrant/licensee, or to have mixed or commingled
        fertilizer from two or more suppliers such that the result
        of either alteration changes the analysis of the fertilizer as
        originally guaranteed, then that person shall become respon-
        sible for obtaining a registration/license and shall be held
        liable for all penalty payments and be subject to other
        provisions of this Act, including seizure, condemnation, and
        stop sale.

(d)     A deficiency in an official sample of mixed fertilizer resulting
        from non-uniformity is not distinguishable from a deficiency
        due to actual plant nutrient shortage and is properly subject
        official action. (Official 1985.)

(e)     Nothing contained in this section shall prevent any person
        from appealing to a court of competent jurisdiction praying
        for judgment as to the justification of such penalty payments.

Section 10. Commercial Value

*For the purpose of determining the commercial value to be applied under the provisions of Section 9 the _____ shall determine and publish annually the values per unit of nitrogen, available ~~phosphoric acid~~ [phosphate], and soluble potash in fertilizers in this state. If guarantees are provided in Section 3(c)(2), the value shall be per unit of nitrogen, phosphorus, and potassium. The value so determined and published shall be used in determining and assessing penalty payments. (Tentative 1992.)*

Section 11. Misbranding

No person shall distribute misbranded fertilizer. A fertilizer shall be deemed to be misbranded:

(a) If its labeling is false or misleading in any particular.

(b) If it is distributed under the name of another fertilizer product.

(c) If it is not labeled as required in Section 5 of this Act and in accordance with regulations prescribed under this Act.

(d) If it purports to be or is represented as a fertilizer, or is represented as containing a plant nutrient or fertilizer unless such plan nutrient or fertilizer conforms to the definition of identity, if any, prescribed by regulation of the _____ ____ ; in adopting such regulations the _____ shall give due regard to commonly accepted definitions and official fertilizer terms such as those issued by the Association of American Plant Food Control Officials.

Section 12.  Adulteration

No person shall distribute an adulterated fertilizer product. A fertilizer shall be deemed to be adulterated:

(a)     If it contains any deleterious or harmful ingredient in sufficient amount to render it injurious to beneficial plant life when applied in accordance with directions for use on the label, or if adequate warning statements or directions for use which may be necessary to protect plant life are not shown upon the label.

(b)     If its composition falls below or differs from that which it is purported to possess by its labeling.

(c)     If it contains unwanted crop seed or weed seed.

Section 13.  Publications

The _____ shall publish at least annually and in such forms as he may deem proper: (a) Information concerning the distribution of fertilizers, (b) Results of analyses based on official samples of fertilizer distributed within the state as compared with analyses guaranteed under Section 4 and Section 5.

Section 14.  Rules and Regulations

The _____ is authorized to prescribe and, after a public hearing following due public notice, to enforce such rules and regulations relating to investigational allowances, definitions, records, and the distribution and storage of fertilizers as may be necessary to carry into effect the full intent and meaning of this Act. (Official 1990.)

Section 15.  Short Weight

If any fertilizer in the possession of the consumer is found by the _____ to be short in weight, the registrant/licensee of said fertilizer shall, within thirty days after official notice from the _____ submit to the consumer a penalty payment of _____ times the value of the actual shortage.

## Section 16.  Cancellation of Registration/License

The _____ is authorized and empowered to cancel the registration (license of any person) of any brand of fertilizer or to refuse to register any brand of fertilizer (issue a license) as herein provided, upon satisfactory evidence that the registrant/licensee has used fraudulent or deceptive practices in the evasion or attempted evasion of the provisions of this Act or any regulations promulgated thereunder: Provided, That no license or registration shall be revoked or refused until the person (registrant/licensee) shall have been given the opportunity to appear for a hearing by the _____ .

## Section 17.  "Stop Sale" Orders

The _____ may issue and enforce a written or printed "stop sale, use, or removal" order to the owner or custodian of any lot of fertilizer and to hold at a designated place when the _____ finds said fertilizer is being offered or exposed for sale in violation of any of the provisions of this Act until the law has been complied with and said fertilizer is released in writing by the _____ , or said violation has been otherwise legally disposed of by written authority. The _____ shall release the fertilizer so withdrawn when the requirements of the provisions of this Act have been complied with and all costs and expenses incurred in connection with the withdrawal have been paid.

## Section 18.   Seizure, Condemnation and Sale

Any lot of fertilizer not in compliance with the provisions of the Act shall be subject to seizure on complaint of the _____ to a court of competent jurisdiction in the area in which said fertilizer is located. In the event the court finds the said fertilizer to be in violation of this Act and orders the condemnation of said fertilizer, it shall be disposed of in any manner consistent with the quality of the fertilizer and the laws of the state: Provided, That in no instance shall the disposition of said fertilizer be ordered by the court without first giving the claimant an opportunity to apply to the court for release of said fertilizer, or for permission to process or relabel said fertilizer to bring it into compliance with this Act.

## Section 19.   Violations

(a)   If it shall appear from the examination of any fertilizer that any of the provisions of this Act or the rules or regulations issued thereunder have been violated, the _____ shall cause notice of the violations to be given to the registrant/licensee or distributor from whom said sample was taken; any person so notified shall be given opportunity to be heard under such regulations as may be prescribed by the _____ . If it appears after such hearing, either in the presence or absence of the person so notified, that any of the provisions of this Act or rules and regulations issued thereunder have been violated, the _____ may certify the facts to the proper prosecuting attorney. (Official 1989.)

(b)   Any person convicted of violating any provision of this Act, or the rules and regulations issued thereunder, shall be punished in the discretion of the court.

(c)   Nothing in this Act shall be construed as requiring the

_____ or his representative to report for prosecution or for the institution of seizure proceedings as a result of minor violations of the Act when he believes that the public interests will be best served by a suitable notice of warning in writing.

(d) It shall be the duty of each _____ attorney to whom any violation is reported to cause appropriate proceedings to be instituted and prosecuted in a court of competent jurisdiction without delay.

(e) The _____ is hereby authorized to apply for and the court to grant a temporary or permanent injunction restraining any person from violating or continuing to violate any of the provisions of this Act or any rule or regulation promulgated under this Act notwithstanding the existence of other remedies at law. Said injunction to be issued without bond.

Section 20. Cooperation with Other Entities

The _____ may cooperate with and enter into agreement with governmental agencies of this State, other states, and agencies of the Federal Government in order to carry out the purpose and provisions of this Act. (Official 1991.)

Section 21. Exchanges Between Manufacturers

Nothing in this Act shall be construed to restrict or avoid sales or exchanges of fertilizers to each other by importers, manufacturers, or manipulators who mix fertilizer materials for sale, or as preventing the free and unrestricted shipments of fertilizer to manufacturers or manipulators who have registered their brands (are licensed) as required by provisions of this Act.

Section 22.  Constitutionality

If any clause, sentence, paragraph, or part of this Act shall for any reason be judged invalid by any court of competent jurisdiction, such judgment shall not affect, impair, or invalidate the remainder thereof, but shall be confined in its operation to the clause, sentence, paragraph, or part thereof directly involved in the controversy in which such judgment shall have been rendered.

Section 23.  Repeal

All laws and parts of laws in conflict with or inconsistent with the provisions of this Act are hereby repealed.

Section 24.  Effective Date

This Act shall take effect and be in force from and after the first day of _____ .

# FERTILIZER
# RULES AND REGULATIONS

Under the Uniform Fertilizer Bill by the _____ of the State of _____ . Pursuant to due publication and notice of opportunity for a public hearing, the _____ has adopted the following regulations.

1.  Plant Nutrients in Addition to Nitrogen, Phosphorus, and Potassium.

    Other plant nutrients when mentioned in any form or manner shall be registered and shall be guaranteed. Guarantees shall be made on the elemental basis. Sources of the elements guaranteed and proof of availability shall be provided the upon request. Except guarantees for those water soluble nutrients labeled for hydroponic or continuous liquid feed programs, the minimum percentages which will be accepted for registration are as follows:

| Element | % |
|---|---|
| Calcium (Ca) | 1.0000 |
| Magnesium (Mg) | 0.5000 |
| Sulfur (S) | 1.0000 |
| Boron (B) | 0.0200 |
| Chlorine (Cl) | 0.1000 |
| Cobalt (Co) | 0.0005 |
| Copper (Cu) | 0.0500 |
| Iron (Fe) | 0.1000 |
| Manganese (Mn) | 0.0500 |
| Molybdenum (Mo) | 0.0005 |
| Sodium (Na) | 0.1000 |
| Zinc (Zn) | 0.0500 |

Guarantees or claims for the above listed plant nutrients are the only ones which will be accepted. Proposed labels and directions for the use of the fertilizer shall be furnished with the application for registration upon request. Any of the above listed elements which are guaranteed shall appear in the order listed immediately following guarantees for the primary nutrients of nitrogen, phosphorus, and potassium. (Official 1987.)

*Secretary's Note - Paragraphs 3 and 4 (Off. Publication No. 38) were deleted - Official 1985.*

A warning or caution statement may be required for any product which contains (name of micro-nutrient) in water soluble form when there is evidence that (name of micro-nutrient) in excess of % may be harmful to certain crops or where there are unusual environmental conditions. (Official 1984.)

## Examples of Warning or Caution Statements:

1.  Directions: Apply this fertilizer at a maximum rate of (number of pounds) per acre for (name of crop).

    CAUTION: Do not use on other crops. The (name of micro-nutrient) may cause injury to them.

2.  CAUTION: Apply this fertilizer at a maximum rate of (number of pounds) per acre for (name of crop). Do not use on other crops; the (name of micro-nutrient) may cause serious injury to them.

3.  WARNING: This fertilizer carries added (name of micro-nutrient) and is intended for use only on (name of crop). Its use on any other crops or under conditions other than those recommended may result in serious injury to the crops.

4.  CAUTION: This fertilizer is to be used only on soil which responds to (name of micro-nutrient). Crops high in (name of micro-nutrient) are toxic to grazing animals (ruminants). (Official 1991.)

*Secretary's Note - Example Warning and Caution statements for boron and molybdenum (page 36 Off. Pub. No. 43) were deleted and above generic statements substituted - Official 1991.*

2.  Fertilizer Labels.

    The following information, in the format presented, is the minimum required for all fertilizer labels. For packaged products, this information shall either (1) appear on the front or back of the package, (2) occupy at least the upper-third of a side of the package, or (3) be printed on a tag and attached to the package. This information shall be in a readable and conspicuous form. For bulk products, this same information in written or printed form shall accompany delivery and be supplied to the purchaser at time of delivery.

(a)  Net weight

(b)  Brand

(c)  Grade (Provided that the grade shall not be required when no primary nutrients are claimed.)

*(d)*  *Guaranteed Analysis\**

*Total Nitrogen (N)\*\** . . . . . . _____ %

_____ % **Ammoniacal Nitrogen**

_____ % **Nitrate Nitrogen**

_____ % **Water Insoluble Nitrogen**

_____ % **Urea Nitrogen**

_____ % **(Other recognized and determinable forms of N)**

*Available* ~~*Phosphoric Acid*~~

*[Phosphate](P$_2$O$_5$)* . . . . . . . _____ %

*Soluble Potash (K$_2$O)* . . . . . . _____ %

*(Other nutrients, elemental basis)\*\*\**_____ %

*(Tentative 1992.)*

(e)  Sources of nutrients, when shown on the label, shall be listed below the completed guaranteed analysis statement.

(f)  Name and address of registrant or licensee.

\*    *Zero (0) guarantees should not be made and shall not appear in statement [except in nutrient guarantee breakdowns]. (Tentative 1992.)*

\*\*   If chemical forms of N are claimed or required, the form shall be shown and the percentages of the individual forms shall add up to the total nitrogen percentage. No implied order of the forms of nitrogen is intended. (Official 1992.)

*** As prescribed by regulation No. 1.
(Official 1986.)

3.  Slowly Released Plant Nutrients

(a) No fertilizer label shall bear a statement that connotes or implies that certain plant nutrients contained in a fertilizer are released slowly over period of time, unless the slow release components are identified and guaranteed at a level of at least 15% of the total guarantee for that nutrient(s). (Official 1991)

(b) Types of products with slow release properties recognized are (1) water insoluble, such as natural organics, ureaform materials, urea-formaldehyde products, isobutylidene diurea, oxamide, etc., (2) coated slow release, such as sulfur coated urea and other encapsulated soluble fertilizers, (3) occluded slow release, where fertilizers or fertilizer materials are mixed with waxes, resins, or other inert materials and formed into particles, and (4) products containing water soluble nitrogen such as ureaform materials, urea-formaldehyde products, methylenediurea (MDU, dimethylenetriurea (DMTU), dicyanodiamide (DCD), etc. The terms, "water insoluble," "coated slow release," "slow release," "controlled release," "slowly available water soluble," and "occluded slow release" are accepted as descriptive of these products, provided the manufacturer can show a testing program substantiating the claim (testing under guidance of Experiment Station personnel or a recognized reputable researcher acceptable to the ___
___ ). A laboratory procedure, acceptable to the _____ , for evaluating the release

characteristics of the product(s) must also be provided by the manufacturer. (Official 1991.)

(c) To supplement (a) and (b), if an amount of nitrogen is designated as organic, then the water insoluble nitrogen or the slow release nitrogen guarantee must not be less than 60% of the nitrogen so designated. Coated urea shall not be included in meeting the 60% requirement. (Official 1992.)

(d) Until more appropriate methods are developed, AOAC method 970.04 (15th Edition) is to be used to confirm the coated slow release and occluded slow release nutrients and others whose slow release characteristics depend on particle size. AOAC method 945.01 (15th Edition) shall be used to determine the water insoluble nitrogen of organic materials. (Official 1991.)

4. Definitions.

Except as the _____ designates otherwise in specific cases, the names and definitions for commercial fertilizers shall be those adopted by the Association of American Plant Food Control Officials.

5. Percentages.

The term of "percentage" by symbol or word, when used on a fertilizer label, shall represent only the amount of individual plant nutrients in relation to the total product by weight.

6. Investigational Allowances.

(a) *A commercial fertilizer shall be deemed deficient if the analysis of any nutrient is below the guarantee by an amount exceeding the values* *in the following schedule, or if the overall index value of the fertilizer is below 98%\*.*

| Guaranteed percent | Nitrogen percent | Available ~~Phosphoric Acid~~ [Phosphate], percent | Potash percent |
|---|---|---|---|
| 04  or  less | 0.49 | 0.67 | 0.41 |
| 05 | 0.51 | 0.67 | 0.43 |
| 06 | 0.52 | 0.67 | 0.47 |
| 07 | 0.54 | 0.68 | 0.53 |
| 08 | 0.55 | 0.68 | 0.60 |
| 09 | 0.57 | 0.68 | 0.65 |
| 10 | 0.58 | 0.69 | 0.70 |
| 12 | 0.61 | 0.69 | 0.79 |
| 14 | 0.63 | 0.70 | 0.87 |
| 16 | 0.67 | 0.70 | 0.94 |
| 18 | 0.70 | 0.71 | 1.01 |
| 20 | 0.73 | 0.72 | 1.08 |
| 22 | 0.75 | 0.72 | 1.15 |
| 24 | 0.78 | 0.73 | 1.21 |
| 26 | 0.81 | 0.73 | 1.27 |
| 28 | 0.83 | 0.74 | 1.33 |
| 30 | 0.86 | 0.75 | 1.39 |
| 32  or  more | 0.88 | 0.76 | 1.44 |

*(Tentative  1992.)*

For guarantees not listed, calculate the appropriate value by interpolation.

---

\*     For these investigational allowances to be applicable, the recommended AOAC procedures for
obtaining samples, preparation, and analysis must
be used. These are described in Official Methods
of Analysis of the Association of Official Analytical
Chemists, 13th Edition, 1980, and in succeeding
issues of the Journal of the Association of Official
Analytical Chemists. In evaluating replicate data,

Table 19, page 935, Journal of the Association
of Official Analytical Chemists, Volume 49, No. 5,
October 1966, should be followed.

The overall index value is calculated by comparing the
commercial value guaranteed with the commercial value
found. Unit values of the nutrients used shall be those
referred to in Section 10 of the Act.

*Overall index value - Example of calculation for
a 10-10-10 grade found to contain 10.1% Total
Nitrogen (N), 10.2% Available* ~~Phosphoric Acid~~
*[Phosphate] ($P_2O_5$), and 10.1% Soluble Potash
($K_2O$). Nutrient unit values are assumed to be
$3 per unit N, $2 per unit $P_2O_5$, and $1 per
unit $K_2O$. (Tentative 1992.)*

| | | |
|---|---|---|
| 10.0 units N | ×3= | 30.0 |
| 10.0 units $P_2O_5$ | ×2= | 20.0 |
| 10.0 units $K_2O$ | ×1= | 10.0 |
| Commercial Value Guaranteed | = | 60.0 |
| | | |
| 10.1 units N | ×3= | 30.3 |
| 10.2 units $P_2O_5$ | ×2= | 20.4 |
| 10.1 units $K_2O$ | ×1= | 10.1 |
| Commercial Value Found | = | 60.8 |

Overall Index Value = 100(60.8/60.00) =101.3%

(b) Secondary and minor elements shall be deemed deficient
if any element is below the guarantee by an amount
exceeding the values in the following schedule:

| Element | | Investigational Allowance |
|---|---|---|
| Calcium | ) | 0.2 unit + 5% of guarantee |
| Magnesium | ) | 0.2 unit + 5% of guarantee |
| Sulfur | ) | 0.2 unit + 5% of guarantee |
| Boron | ) | 0.003 unit + 15% of guarantee |

| Cobalt | ) | 0.0001 unit + 30% of guarantee |
| Molybdenum | ) | 0.0001 unit + 30% of guarantee |
| Chlorine | ) | 0.005 unit + 10% of guarantee |
| Copper | ) | 0.005 unit + 10% of guarantee |
| Iron | ) | 0.005 unit + 10% of guarantee |
| Manganese | ) | 0.005 unit + 10% of guarantee |
| Sodium | ) | 0.005 unit + 10% of guarantee |
| Zinc | ) | 0.005 unit + 10% of guarantee |

The maximum allowance when calculate in accordance to the above shall be 1 unit (1%).

7.   Sampling.

Sampling equipment and procedures shall be those adopted by the Association of Official Analytical Chemists wherever applicable.

8.   Breakdown of Plant Food Elements Within the Guaranteed Analysis.

When a plant nutrient guarantee is broken down into the component forms, the percentage for each component shall be shown before the name of the form.

EXAMPLES:

Total Nitrogen (N) . . . . . . . . . . . . . _____ %

_____ % Ammoniacal Nitrogen

_____ % Nitrate Nitrogen

Magnesium (Mg) . . . . . . . . . . . . . _____ %

_____ % Water Soluble Magnesium (Mg)

Sulfur (S) . . . . . . . . . . . . . . _____ %

_____ % Free Sulfur (S)

_____ % Combined Sulfur (S)

Iron (Fe) %

_____ % Chelated Iron (Fe)

Manganese (Mn) . . . . . . . . . . . _____ %

_____ % Water Soluble Manganese (Mn)

(Official 1992.)

# Appendix B

# *Useful Tables and Conversions*

## Table B-1
## Conversion Factors for U.S. and Metric Units

| To Convert Column 1 into Column 2, Multiply by | Column 1 | Column 2 | To Convert Column 2 into Column 1, Multiply by |
|---|---|---|---|
| *Length* | | | |
| 0.621 | kilometer, km | mile, mi. | 1.609 |
| 1.094 | meter, m | yard, yd. | 0.914 |
| 0.394 | centimeter, cm | inch, in. | 2.54 |
| *Area* | | | |
| 0.386 | kilometer$^2$, km$^2$ | mile$^2$, mi.$^2$ | 2.590 |
| 247.1 | kilometer$^2$, km$^2$ | acre, A | 0.00405 |
| 2.471 | hectare, ha | acre, A | 0.405 |
| *Volume* | | | |
| 0.00973 | meter$^3$, m$^3$ | acre-inch | 102.8 |
| 3.532 | hectoliter, hl | cubic foot, ft.$^3$ | 0.2832 |
| 2.838 | hectoliter, hl | bushel, bu. | 0.352 |
| 0.0284 | liter, L | bushel, bu. | 35.24 |
| 1.057 | liter, L | quart (liquid), qt. | 0.946 |
| *Mass* | | | |
| 1.102 | tonne (metric) | ton (short) | 0.9072 |
| 2.205 | kilogram, kg | pound, lb. | 0.454 |
| 0.035 | gram, g | ounce (avdp.), oz. | 28.35 |

*(Continued)*

## Table B-1 (Continued)

| To Convert Column 1 into Column 2, Multiply by | Column 1 | Column 2 | To Convert Column 2 into Column 1, Multiply by |
|---|---|---|---|
| | | *Pressure* | |
| 14.50 | bar | lb./inch², psi | 0.06895 |
| 0.9869 | bar | atmosphere, atm | 1.013 |
| 0.9678 | kg/cm² | atmosphere, atm | 1.033 |
| 14.22 | kg/cm² | lb./inch², psi | 0.07031 |
| 14.70 | atmosphere, atm | lb./inch², psi | 0.06805 |
| 0.1450 | kilopascal | lb./inch², psi | 6.895 |
| 0.009869 | kilopascal | atmosphere, atm | 101.30 |
| | | *Yield or Rate* | |
| 0.446 | tonne (metric)/ hectare | ton (short)/acre | 2.240 |
| 0.891 | kg/ha | lb./acre | 1.12 |

## Table B-2
## Metric-U.S. System Equivalents

*Length*

| Metric denominations and values | | | Equivalents in denominations in use |
|---|---|---|---|
| myriameter .... | = | 10,000 m ....... = | 6.2137 mi. |
| kilometer ...... | = | 1,000 m ....... = | 0.62137 mi. or 3,280 ft 10 in. |
| hectometer ..... | = | 100 m ....... = | 328 ft. 1 in. |
| dekameter ..... | = | 10 m ....... = | 393.7 in. |
| meter ......... | = | 1m ....... = | 39.37 in. |
| decimeter ...... | = | 0.1 m ....... = | 3.937 in. |
| centimeter ..... | = | 0.01 m ....... = | 0.3937 in. |
| millimeter ...... | = | 0.001 m ....... = | 0.0394 in. |

*(Continued)*

## Table B–2 (Continued)

*Volume*

| Name | No. liters | Cubic measure | Dry measure | Liquid measure |
|---|---|---|---|---|
| kiloliter ... = | 1,000 | ... = 1 cu m ... | = 1.308 cu yds.  . | = 264.17 gal. |
| hectoliter.. = | 100 | ... = 0.1 cu m .. | = 2 bu. 3.35 pks. | = 26.417 gal. |
| dekaliter... = | 10 | ... = 10 cu dm .. | = 9.08 qt. ...... | = 2.6417 gal. |
| liter ...... = | 1 | ... = 1 cu dm ... | = 0.908 qt. ..... | = 1.0567 qt. |
| deciliter ... = | 0.1 | ... = 0.1 cu dm . | = 6.1022 cu in. .. | = 0.845 gill |
| centiliter... = | 0.01 | ... = 10 cu cm .. | = 0.6102 cu in. .. | = 0.338 fl. oz. |
| milliliter ... = | 0.001 | ... = 1 cu cm ... | = 0.061 cu in. ... | = 0.27 fluid dr. |

*Weight*

| Name | No. grams | Cubic measure[1] | Avoirdupois weight |
|---|---|---|---|
| millier or tonneau ... = | 1,000,000..... | = 1 cu m ...... | = 2,204.6 lbs. |
| quintal............ = | 100,000..... | = 1 hl ......... | = 220.46 lbs. |
| myriagram......... = | 10,000..... | = 10 L......... | = 22.046 lbs. |
| kilogram or kilo ..... = | 1,000..... | = 1 L.......... | = 2.2046 lbs. |
| hectogram.......... = | 100..... | = 1 dl ......... | = 3.5274 oz. |
| dekagram .......... = | 10..... | = 10 cu cm..... | = 0.3527 oz. |
| gram .............. = | 1..... | = 1 cu cm ..... | = 15.432 gr. |
| decigram........... = | 0.1..... | = 0.1 cu cm .... | = 1.5432 gr. |
| centigram .......... = | 0.01..... | = 10 cu mm.... | = 0.1543 gr. |
| milligram........... = | 0.001..... | = 1 cu mm..... | = 0.0154 gr. |

*Area*

| hectare................ = | 10,000 sq. m............. | = 2.471 acres |
|---|---|---|
| are ................... = | 100 sq. m............. | = 119.6 sq. yds. |
| centare................ = | 1 sq. m............. | = 1,550 sq. in. |

[1]Based on pure water at 4°C and 760 mm pressure.

## Table B–3
## Temperature Comparison of Celsius to Fahrenheit

| Celsius (C°) | | Fahrenheit (F°) |
|---|---|---|
| –30 | . . . . . . . . . . . . . . . . | –22 |
| –20 | . . . . . . . . . . . . . . . . | – 4 |
| –10 | . . . . . . . . . . . . . . . . | 14 |
| 0 | . . . . . . . . . . . . . . . . | 32 |
| 10 | . . . . . . . . . . . . . . . . | 50 |
| 20 | . . . . . . . . . . . . . . . . | 68 |
| 30 | . . . . . . . . . . . . . . . . | 86 |
| 40 | . . . . . . . . . . . . . . . . | 104 |
| 50 | . . . . . . . . . . . . . . . . | 122 |
| 60 | . . . . . . . . . . . . . . . . | 140 |
| 70 | . . . . . . . . . . . . . . . . | 158 |
| 80 | . . . . . . . . . . . . . . . . | 176 |
| 90 | . . . . . . . . . . . . . . . . | 194 |
| 100 | . . . . . . . . . . . . . . . . | 212 |

*Conversion Formulas*

$$C° = \frac{5}{9}(F° - 32) \qquad F° = (\frac{9}{5} C°) + 32$$

## Table B–4
## Useful Conversions

The following data are useful in calculating rates of application:

| | |
|---|---|
| 1 acre-foot of soil | = 4,000,000 lbs. (approximate) |
| 1 t. per acre | = 20.8 g per sq. ft. |
| 1 t. per acre | = 1 lb. per 21.78 sq. ft. |
| 1 t. per acre | = 25.12 quintals per hectare |
| 1 t. per acre 6" depth | = 1 g per 1,000 g of soil |
| 1 g per sq. ft. | = 96 lbs. per acre |
| 1 lb. per acre | = 0.0104 g per sq. ft. |
| 1 lb. per acre | = 1.12 kilos per hectare |
| 100 lbs. per acre | = 0.2296 lbs. per 100 sq. ft. |
| g per sq. ft. × 96 | = lbs. per acre |
| kg per 48 sq. ft. | = t. per acre |
| lbs. per sq. ft. × 21.78 | = t. per acre |
| lbs. per sq. ft. × 43,560 | = lbs. per acre |
| 100 sq. ft. | = 1/435.6 or 0.002296 acre |
| t. per acre-foot | = 0.00136 × parts per million |
| parts per million | = 17.1 × gr. per gal. |
| parts per million × 0.00136 | = t. per acre-foot |

### Table B–5
### Useful Weights and Measures for the
### Home Gardener and Horticulturist

*Weights*

| Pounds per Acre | | Equivalent Quantity per 100 Square Feet |
|---|---|---|
| 100 | . . . . . . . . . . . . . . . | 3$\frac{1}{2}$ oz. |
| 200 | . . . . . . . . . . . . . . . | 7$\frac{1}{2}$ oz. |
| 300 | . . . . . . . . . . . . . . . | 11 oz. |
| 400 | . . . . . . . . . . . . . . . | 14$\frac{3}{4}$ oz. |
| 500 | . . . . . . . . . . . . . . . | 1 lb. 2$\frac{1}{2}$ oz. |
| 600 | . . . . . . . . . . . . . . . | 1 lb. 6 oz. |
| 700 | . . . . . . . . . . . . . . . | 1 lb. 10 oz. |
| 800 | . . . . . . . . . . . . . . . | 1 lb. 13 oz. |
| 900 | . . . . . . . . . . . . . . . | 2 lbs. 1 oz. |
| 1,000 | . . . . . . . . . . . . . . . | 2 lbs. 5 oz. |
| 2,000 | . . . . . . . . . . . . . . . | 4 lbs. 10 oz. |

*Measures*
*(approximate)*

| | | |
|---|---|---|
| 1 level tsp. | . . . . . . . . . . . . . . . | $\frac{1}{6}$ oz. |
| 1 level tbsp. | . . . . . . . . . . . . . . . | $\frac{1}{2}$ oz. |
| 1 level c. | . . . . . . . . . . . . . . . | 8 oz. |
| 1 pt. | . . . . . . . . . . . . . . . | 1 lb. |
| 1 qt. | . . . . . . . . . . . . . . . | 2 lbs. |
| 1 gal. | . . . . . . . . . . . . . . . | 8 lbs. |

## Table B–6
## Land Measurements

| Linear Measure | | Square Measure | |
|---|---|---|---|
| 1 in. | 0.0833 ft. | 144 sq. in. | 1 sq. ft. |
| 7.92 in. | 1 link | 9 sq. ft. | 1 sq. yd. |
| 12 in. | 1 ft. | $30\frac{1}{4}$ sq. yds. | 1 sq. rd. |
| 1 vara | 33 in. | 16 sq. rds. | 1 sq. ch. |
| $2\frac{3}{4}$ ft. | 1 yard | | |
| 3 ft. | 1 yd. | 1 sq. ch. | 4,356 sq. ft. |
| 25 links | $16\frac{1}{2}$ ft. | 10 sq. chs. | 1 acre |
| 25 links | 1 rd. | 160 sq. rds. | 1 acre |
| 100 links | 1 ch. | 4,840 sq. yds. | 1 acre |
| $16\frac{1}{2}$ ft. | 1 rod | | |
| $5\frac{1}{2}$ yd. | 1 rod | | |
| 4 rds. | 100 links | 1 sq. mi. | 1 section |
| 66 ft. | 1 ch. | 160 acres | $\frac{1}{4}$ section |
| 80 ch. | 1 mi. | 36 sq. mi. | 1 twp. |
| 320 rds. | 1 mi. | 6 mi. sq. | 1 twp. |
| 8,000 links | 1 mi. | 1 sq. mi. | 2.59 sq. km |
| 5,280 ft. | 1 mi. | | |
| 1,760 yds. | 1 mi. | | |
| 9 in. | 1 span | | |
| 4 in. | 1 hand | | |

## Table B–7
## Volume Measurements

*Cubic Measure*

1,728 cu in. . . . . . . . . . . . . . . . . . . . . . . . . . . 1 cu ft.

1 cu ft. . . . . . . . . . . . . . . . . . . . . . . . . . . . . . 7.4805 gal.

27 cu ft. . . . . . . . . . . . . . . . . . . . . . . . . . . . . 1 cu yd.

128 cu ft. (4′ × 4′ × 8′) . . . . . . . . . . . . . . . 1 cord (wood)

231 cu in. . . . . . . . . . . . . . . . . . . . . . . . . . . 1 gal.

2,150.4 cu in. . . . . . . . . . . . . . . . . . . . . . . 1 bu.

1.244 cu ft. . . . . . . . . . . . . . . . . . . . . . . . . 1 bu.

*Liquid Measure*

1 pt. (4 gills) . . . . . . . . . . . . . . . . . . . . . . . 16 fluid oz.

1 qt. (2 pts.) . . . . . . . . . . . . . . . . . . . . . . . . 32 fluid oz.

1 gal. (4 qt.) . . . . . . . . . . . . . . . . . . . . . . . . 128 fluid oz.

1 gal. (U.S.) . . . . . . . . . . . . . . . . . . . . . . . . 0.8327 imperial gal.

$31\frac{1}{2}$ gal. . . . . . . . . . . . . . . . . . . . . . . . . . . . . 1 barrel

42 gal. . . . . . . . . . . . . . . . . . . . . . . . . . . . . 1 barrel (petroleum measure)

63 gal. (2 barrels) . . . . . . . . . . . . . . . . . . . 1 hogshead

*Dry Measure*

2 pts. dry . . . . . . . . . . . . . . . . . . . . . . . . . . 1 qt. dry

8 qt. dry . . . . . . . . . . . . . . . . . . . . . . . . . . . 1 pk.

4 pks. . . . . . . . . . . . . . . . . . . . . . . . . . . . . . 1 bu.

105 qt. dry or 7,056 cu in. . . . . . . . . . . . . . 1 standard barrel

*Gallons in Square Tanks*

To find the number of gallons in a square or oblong tank, multiply the number of cubic feet that it contains by 7.4805.

*Gallons in Circular Tanks*

To find the number of gallons in a circular tank or well, square the diameter in feet, multiply by the depth in feet, and then multiply by 5.875

Table B–8
Convenient Conversion Factors

| Multiply | By | To Get |
|---|---|---|
| Acres | 0.4048 | Hectares |
| Acres | 43,560 | Square feet |
| Acres | 160 | Square rods |
| Acres | 4,840 | Square yards |
| Ares | 1,076.4 | Square feet |
| Bushels | 4 | Pecks |
| Bushels | 64 | Pints |
| Bushels | 32 | Quarts |
| Centimeters | 0.3937 | Inches |
| Centimeters | 0.01 | Meters |
| Cubic centimeters | 0.03382 | Ounces (liquid) |
| Cubic feet | 1,728 | Cubic inches |
| Cubic feet | 0.03704 | Cubic yards |
| Cubic feet | 7.4805 | Gallons |
| Cubic feet | 29.92 | Quarts (liquid) |
| Cubic yards | 27 | Cubic feet |
| Cubic yards | 46,656 | Cubic inches |
| Cubic yards | 202 | Gallons |
| Feet | 30.48 | Centimeters |
| Feet | 12 | Inches |
| Feet | 0.3048 | Meters |
| Feet | 0.060606 | Rods |
| Feet | $\frac{1}{3}$ or 0.33333 | Yards |
| Feet per minute | 0.01136 | Miles per hour |
| Gallons | 0.1337 | Cubic feet |
| Gallons | 4 | Quarts (liquid) |
| Gallons of water | 8.3453 | Pounds of water |
| Grams | 15.43 | Grains |
| Grams | 0.001 | Kilograms |
| Grams | 1,000 | Milligrams |
| Grams | 0.0353 | Ounces |
| Grams per liter | 1,000 | Parts per million |
| Hectares | 2.471 | Acres |
| Inches | 2.54 | Centimeters |
| Inches | 0.08333 | Feet |

*(Continued)*

## Table B–8 (Continued)

| Multiply | By | To Get |
|---|---|---|
| Kilograms | 1,000 | Grams |
| Kilograms | 2.205 | Pounds |
| Kilograms per hectare | 0.892 | Pounds per acre |
| Kilometers | 3,281 | Feet |
| Kilometers | 0.6214 | Miles |
| Liters | 1,000 | Cubic centimeters |
| Liters | 0.0353 | Cubic feet |
| Liters | 61.02 | Cubic inches |
| Liters | 0.2642 | Gallons |
| Liters | 1.057 | Quarts (liquid) |
| Meters | 100 | Centimeters |
| Meters | 3.2181 | Feet |
| Meters | 39.37 | Inches |
| Miles | 5,280 | Feet |
| Miles | 63,360 | Inches |
| Miles | 320 | Rods |
| Miles | 1,760 | Yards |
| Miles per hour | 88 | Feet per minute |
| Miles per hour | 1.467 | Feet per second |
| Miles per minute | 60 | Miles per hour |
| Ounces (dry) | 0.0625 | Pounds |
| Ounces (liquid) | 0.0625 | Pints (liquid) |
| Ounces (liquid) | 0.03125 | Quarts (liquid) |
| Parts per million | 8.345 | Pounds per million gallons of water |
| Pecks | 16 | Pints (dry) |
| Pecks | 8 | Quarts (dry) |
| Pints (dry) | 0.5 | Quarts (dry) |
| Pints (liquid) | 16 | Ounces (liquid) |
| Pounds | 453.5924 | Grams |
| Pounds | 16 | Ounces |
| Pounds of water | 0.1198 | Gallons |
| Quarts (liquid) | 0.9463 | Liters |
| Quarts (liquid) | 32 | Ounces (liquid) |
| Quarts (liquid) | 2 | Pints (liquid) |
| Rods | 16.5 | Feet |
| Rods | 5.5 | Yards |

*(Continued)*

## Table B-8 (Continued)

| Multiply | By | To Get |
|---|---|---|
| Square feet | 144 | Square inches |
| Square feet | 0.11111 | Square yards |
| Square inches | 0.00694 | Square feet |
| Square miles | 640 | Acres |
| Square miles | 27,878,400 | Square feet |
| Square rods | 0.00625 | Acres |
| Square rods | 272.25 | Square feet |
| Square yards | 0.0002066 | Acres |
| Square yards | 9 | Square feet |
| Square yards | 1,296 | Square inches |
| Temperature (°C) + 17.98 | 1.8 | Temperature, °F |
| Temperature (°F) − 32 | $5/9$ or 0.5555 | Temperature, °C |
| Tons | 907.1849 | Kilograms |
| Tons | 2,000 | Pounds |
| Tons, long | 2,240 | Pounds |
| Yards | 3 | Feet |
| Yards | 36 | Inches |
| Yards | 0.9144 | Meters |

## Table B-9
## Test Weights of Agricultural Products[1]

| Commodity | Unit | Approximate New Weight |
|---|---|---|
| | | *(lbs.)* |
| Alfalfa seed | Bu. | 60 |
| Apples | Bu. | 48 |
| | Northwest box | 44 |
| | Eastern box | 54 |
| Apricots | Lug (Campbell) | 24 |
| | 4-basket crate | 26 |
| Barley | Bu. | 48 |
| Beans, dry | Bu. | 60 |
| Beans, lima | Bu. | 56 |
| Beans, snap | Bu. | 30 |
| Bluegrass seed | Bu. | 14–30 |

*(Continued)*

## Table B–9 (Continued)

| Commodity | Unit | Approximate New Weight |
|---|---|---|
| | | *(lbs.)* |
| Canola | Bu. | 50 |
| Cherries | Lug (Campbell) | 16 |
| Clover seed | Bu. | 60 |
| Corn, ear (husked) | Approx. 2 bu. | 70$^2$ |
| Corn, shelled | Bu. | 56 |
| Cotton | Bale, gross | 500 |
| | Bale, net | 480 |
| Cottonseed | Bu. | 32 |
| Flaxseed | Bu. | 56 |
| Grain sorghums | Bu. | 56 |
| Grapefruit | Box (Tex. 1$\frac{2}{5}$-bu.) | 80 |
| | Box (Calif. desert & Ariz.) | 64 |
| Lemons, California | Box (Calif. & Ariz.) | 76 |
| Lentils | Bu. | 60 |
| Limes, Florida | Box | 80 |
| Meadow fescue seed | Bu. | 24 |
| Milk | Gal. | 8.6 |
| Millet | Bu. | 48–60 |
| Milo | Bu. | 56 |
| Mustard seed | Bu. | 58–60 |
| Oats | Bu. | 32 |
| Olives | Lug | 25–30 |
| Onions, dry | Sack | 50 |
| Oranges | Box (Fla. & Tex.) | 90 |
| | Box (Calif. & Ariz.) | 75 |
| Orchardgrass seed | Bu. | 14 |
| Peaches | Lug | 20 |
| | Bu. | 48 |
| Peanuts, unshelled | | |
|   Runners | Bu. | 21 |
|   Spanish | Bu. | 25 |
|   Virginia type | Bu. | 17 |
| Peas, dry | Bu. | 60 |

*(Continued)*

## Table B–9 (Continued)

| Commodity | Unit | Approximate New Weight |
|---|---|---|
| | | *(lbs.)* |
| Potatoes | Bu. | 60 |
| Rice, rough | Bu. | 45 |
| Rye | Bu. | 56 |
| Soybeans | Bu. | 60 |
| Sudangrass seed | Bu. | 40 |
| Sweet potatoes | Bu. | 55 |
| Timothy seed | Bu. | 45 |
| Tomatoes | Crate | 60 |
| | Lug | 32 |
| Turpentine | Gal. | 7.23 |
| Vetch seed | Bu. | 60 |
| Walnuts | Bu. | 50 |
| Wheat | Bu. | 60 |

[1]Taken from *Agricultural Statistics*, published annually by USDA.
[2]Amount needed to give 1 bu. shelled corn.

## Table B–10
## Number of Trees or Plants on an Acre

| Spacing | Number | Spacing | Number |
|---|---|---|---|
| 1 by 2 ft. | 21,780 | 6 by 6 ft. | 1,210 |
| 1 by 3 ft. | 14,520 | 6 by 8 ft. | 907 |
| 1 by 4 ft. | 10,890 | 8 by 8 ft. | 680 |
| 1½ by 2 ft. | 14,520 | 10 by 10 ft. | 436 |
| 1½ by 3 ft. | 9,680 | 12 by 12 ft. | 302 |
| 2 by 3 ft. | 7,260 | 15 by 15 ft. | 194 |
| 2 by 4 ft. | 5,445 | 16 by 16 ft. | 170 |
| 3 by 4 ft. | 3,630 | 18 by 18 ft. | 134 |
| 3 by 5 ft. | 2,904 | 20 by 20 ft. | 109 |
| 3 by 6 ft. | 2,420 | 25 by 25 ft. | 70 |
| 4 by 4 ft. | 2,722 | 30 by 30 ft. | 48 |
| 4 by 6 ft. | 1,815 | 40 by 40 ft. | 27 |

Table B–11
Atomic Weights of Elements in Common Fertilizer Materials

| Element | Symbol | Atomic Weight |
|---|---|---|
| Aluminum | Al | 26.97 |
| Boron | B | 10.82 |
| Calcium | Ca | 40.08 |
| Carbon | C | 12.01 |
| Chlorine | Cl | 35.46 |
| Cobalt | Co | 58.94 |
| Copper | Cu | 63.54 |
| Fluorine | F | 19.00 |
| Hydrogen | H | 1.01 |
| Iodine | I | 126.92 |
| Iron | Fe | 55.85 |
| Magnesium | Mg | 24.31 |
| Manganese | Mn | 54.94 |
| Molybdenum | Mo | 95.94 |
| Nickel | Ni | 58.69 |
| Nitrogen | N | 14.01 |
| Oxygen | O | 16.00 |
| Phosphorus | P | 30.98 |
| Potassium | K | 39.10 |
| Sodium | Na | 23.00 |
| Sulfur | S | 32.06 |
| Zinc | Zn | 65.37 |

Table B–12
Chemical Symbols, Equivalent Weights and Common Names
of Ions, Salts and Chemical Amendments

| Chemical Symbol or Formula | Gram Equivalent Weight | Common Name |
|---|---|---|
| $Ca^{++}$ | 20.04 | Calcium ion |
| $Mg^{++}$ | 12.15 | Magnesium ion |
| $Na^+$ | 23.00 | Sodium ion |
| $K^+$ | 39.10 | Potassium ion |
| $Cl^-$ | 35.46 | Chloride ion |
| $NO_3^-$ | 62.01 | Nitrate ion |
| $NH_4^+$ | 17.03 | Ammonium ion |
| $SO_4^-$ | 48.03 | Sulfate ion |
| $CO_3^-$ | 30.00 | Carbonate ion |
| $HCO_3^-$ | 61.02 | Bicarbonate ion |
| $CaCl_2$ | 55.50 | Calcium chloride |
| $CaSO_4$ | 68.07 | Calcium sulfate |
| $CaSO_4 \bullet 2H_2O$ | 86.09 | Gypsum |
| $CaCO_3$ | 50.04 | Calcium carbonate |
| $MgCl_2$ | 47.62 | Magnesium chloride |
| $MgSO_4$ | 60.19 | Magnesium sulfate |
| $MgCO_3$ | 42.16 | Magnesium carbonate |
| $NaCl$ | 58.46 | Sodium chloride |
| $Na_2SO_4$ | 71.03 | Sodium sulfate |
| $Na_2CO_3$ | 53.00 | Sodium carbonate |
| $NaHCO_3$ | 84.02 | Sodium bicarbonate |
| $KCl$ | 74.56 | Potassium chloride |
| $K_2SO_4$ | 87.13 | Potassium sulfate |
| $K_2S_2O_3$ | 95.16 | Potassium thiosulfate |
| $K_2CO_3$ | 69.10 | Potassium carbonate |
| $KHCO_3$ | 100.12 | Potassium bicarbonate |
| $S$ | 16.03 | Sulfur |
| $SO_2$ | 32.03 | Sulfur dioxide |
| $H_2SO_4$ | 49.04 | Sulfuric acid |
| $Al_2(SO_4)_3 \bullet 18H_2O$ | 111.08 | Aluminum sulfate |
| $FeSO_4 \bullet 7H_2O$ | 139.02 | Iron sulfate (ferrous) |

## Table B-13
## Average Nutrient Analysis of Some Organic Materials

|  | N | $P_2O_5$ | $K_2O$ |
|---|---|---|---|
|  | - - - - - - - - - - - (%) - - - - - - - - - - - | | |
| *Fresh manure with normal quantity of bedding or litter* | | | |
| Duck | 1.1 | 1.45 | 0.50 |
| Goose | 1.1 | 0.55 | 0.50 |
| Turkey | 1.3 | 0.70 | 0.50 |
| Rabbit | 2.0 | 1.33 | 1.20 |
| *Bulky organic materials* | | | |
| Alfalfa hay | 2.5 | 0.50 | 2.10 |
| Bean straw | 1.2 | 0.25 | 1.25 |
| Grain straw | 0.6 | 0.20 | 1.10 |
| Cotton gin trash | 0.7 | 0.18 | 1.19 |
| Seaweed (kelp) | 0.2 | 0.10 | 0.60 |
| Winery pomace (dried) | 1.5 | 1.50 | 0.75 |
| *Organic concentrates* | | | |
| Dried blood | 12.0 | 1.5 | — |
| Fish meal | 10.4 | 5.9 | — |
| Digested sewage sludge | 2.0 | 3.0 | — |
| Activated sewage sludge | 6.5 | 3.4 | 0.3 |
| Tankage | 7.0 | 8.6 | 1.5 |
| Cottonseed meal | 6.5 | 3.0 | 1.5 |
| Bat guano | 13.0 | 5.0 | 2.0 |
| Bone meal[1] | <1.0 | 12–14 | — |

[1]Bone meal values vary widely because of moisture content and processing. Available $P_2O_5$, 12%–14%; insoluble $P_2O_5$, 14%–16%; total $P_2O_5$, 26%–28%.

### Table B–14
### The Neutralizing Value of the Pure Forms
### of Commonly Used Liming Materials

| Material | Chemical | Neutralizing Value |
|---|---|---|
| | | *(%)* |
| Calcium oxide | CaO | 179 |
| Calcium hydroxide | $Ca(OH)_2$ | 136 |
| Dolomite | $CaMg(CO_3)_2$ | 109 |
| Calcium carbonate | $CaCO_3$ | 100 |
| Calcium silicate | $CaSiO_3$ | 86 |

Use of the neutralizing value makes possible the most simple and straight-forward comparison of one liming material with another.

### Table B–15
### Salt Index (Relative Effect of Fertilizer Materials
### on the Soil Solution)[1]

| Material | Salt Index | Partial Salt Index per Unit of Plant Nutrient |
|---|---|---|
| Anhydrous ammonia | 47.1 | 0.572 |
| Ammonium nitrate | 104.7 | 2.990 |
| Ammonium nitrate–lime | 61.1 | 2.982 |
| Ammonium phosphate (11-48-0) | 26.9 | 2.442 |
| Ammonium polysulfide (20-0-0-40S) | 43.6 | 2.180 |
| Ammonium sulfate | 69.0 | 3.253 |
| Ammonium thiosulfate (12-0-0-26S) | 84.4 | 7.040 |
| Calcium carbonate (limestone) | 4.7 | 0.083 |
| Calcium cyanamide | 31.0 | 1.476 |
| Calcium nitrate | 52.5 | 4.409 |
| Calcium sulfate (gypsum) | 8.1 | 0.247 |
| Diammonium phosphate | 29.9 | 1.614 (N) 0.637 ($P_2O_5$) |

*(Continued)*

## Table B–15 (Continued)

| Material | Salt Index | Partial Salt Index per Unit of Plant Nutrient |
|---|---|---|
| Dolomite (calcium and magnesium carbonates) | 0.8 | 0.042 |
| Kainit, 13.5% | 105.9 | 8.475 |
| Kainit, 17.5% | 109.4 | 6.253 |
| Manure salts, 20% | 112.7 | 5.636 |
| Manure salts, 30% | 91.9 | 3.067 |
| Monoammonium phosphate | 34.2 | 2.453 (N) |
|  |  | 0.485 ($P_2O_5$) |
| Monocalcium phosphate | 15.4 | 0.274 |
| Nitrate of soda | 100.0 | 6.060 (N) |
| Nitrogen solution, 37% | 77.8 | 2.104 |
| Nitrogen solution, 40% | 70.4 | 1.724 |
| Potassium chloride, 50% | 109.4 | 2.189 |
| Potassium chloride, 60% | 116.3 | 1.936 |
| Potassium chloride, 63% | 114.3 | 1.812 |
| Potassium nitrate | 73.6 | 5.336 (N) |
|  |  | 1.580 ($K_2O$) |
| Potassium polysulfide (0-0-22-23S) | 37.7 | 1.720 |
| Potassium sulfate | 46.1 | 0.853 ($K_2O$) |
| Potassium thiosulfate (0-0-25-17S) | 64.0 | 2.560 |
| Sodium chloride | 153.8 | 2.899 (Na) |
| Sulfate of potash—magnesia | 43.2 | 1.971 ($K_2O$) |
| Superphosphate, 16% | 7.8 | 0.487 |
| Superphosphate, 20% | 7.8 | 0.390 |
| Superphosphate, 45% | 10.1 | 0.224 |
| Superphosphate, 48% | 10.1 | 0.210 |
| Urea | 75.4 | 1.618 |
| Urea-ammonium nitrate solution (32-0-0) | 95.0 | 2.304 |

[1]After L. F. Rader, Jr., et al., *Soil Sci.*, 55:21--218, 1943.

## Table B–16
## Units of Water Measurement

*Volume units*

One acre-inch

= 3,630 cu ft.

= 27,154 gal.

= $\frac{1}{12}$üîç,A1çA

One acre-foot

= 43,560 cu ft.

= 325,851 gal.

= 12 acre-inches

One cubic foot

= 1,728 cu in.

= 7.481 (approximately 7.5) gal.

weighs approximately 62.4 lbs. (62.5 for most calculations)

One gallon

= 231 cu in.

= 0.13368 cu ft.

weighs approximately 8.33 lbs.

*Flow units*

One cubic foot per second

= 448.83 (approximately 450) gal. per minute

= 1 acre-inch in 1 hour and 30 seconds (approximately 1 hour), or 0.992 (approximately 1) acre-inch per hour

= 1 acre-foot in 12 hours and 6 minutes (approximately 12 hours), or 1.984 (approximately 2) acre-feet per 24 hours

One gallon per minute

= 0.00223 (approximately $\frac{1}{450}$) cu ft. per second

= 1 acre-inch in 452.6 (approximately 450) hours, or 0.00221 acre-inch per hour

= 1 acre-foot in 226.3 days, or 0.00442 acre-foot per day

= 1 in. depth of water over 96.3 sq. ft. in 1 hour

Million gallons per day

= 1.547 cu ft. per second

= 694.4 gal. per minute

## Table B-17
### Conversion Table for Units of Flow[1]

| Units | Cubic Feet per Second | Gallons per Minute | Million Gallons per Day | Acre-Inches per 24 Hours | Acre-Feet per 24 Hours |
|---|---|---|---|---|---|
| Cu ft. per second | 1.0 | 448.8 | 0.646 | 23.80 | 1.984 |
| Gal. per minute | 0.00223 | 1.0 | 0.00144 | 0.053 | 0.00442 |
| Million gal. per day | 1.547 | 694.4 | 1.0 | 36.84 | 3.07 |
| Acre-inches per 24 hours | 0.042 | 18.86 | 0.0271 | 1.0 | 0.0833 |
| Acre-feet per 24 hours | 0.504 | 226.3 | 0.3259 | 12.0 | 1.0 |

[1]The following approximate formulae may be conveniently used to compute the depth of water applied to a field:

$$\frac{\text{cu. ft. per second} \times \text{hours}}{\text{acres}} = \text{acre-inches per acre, or average depth in in.}$$

$$\frac{\text{gal. per minute} \times \text{hours}}{450 \times \text{acres}} = \text{acre-inches per acre, or average depth in in.}$$

## Table B-18
### Approximate Amounts of Water Held by Different Soils

| Soil Texture | Inches of Water Held per Foot of Soil | Max Rate of Irrigation— Inches per Hour, Bare Soil |
|---|---|---|
| Sand | 0.5-0.7 | 0.75 |
| Fine sand | 0.7-0.9 | 0.60 |
| Loamy sand | 0.7-1.1 | 0.50 |
| Loamy fine sand | 0.8-1.2 | 0.45 |
| Sandy loam | 0.8-1.4 | 0.40 |
| Loam | 1.0-1.8 | 0.35 |
| Silt loam | 1.2-1.8 | 0.30 |
| Clay loam | 1.3-2.1 | 0.25 |
| Silty clay | 1.4-2.5 | 0.20 |
| Clay | 1.4-2.4 | 0.15 |

Plants transpire, on the average, from 0.1 to 0.3 in. of rainfall or irrigation water per day.

### Table B–19
### Easy Conversion of Water Analysis of Ions to
### Pounds Material per Acre-Foot of Water

| Ions | Milli-equivalent weight (x meq/l = ppm) (A) | Pounds per Acre-Foot of Water per Milli-equivalent | Conversion Factor[1] (B) | Equivalent to Material in Pounds per Acre-Foot (C) | |
|---|---|---|---|---|---|
| Ca | 20.0 | 54.4 | 6.8 | 136 | $CaCO_3$ |
| Mg | 12.2 | 33.2 | 9.4 | 115 | $MgCO_3$ |
| Mg | 12.2 | 33.2 | 13.5 | 165. | $MgSO_4$ |
| Na | 23.0 | 62.6 | 6.9 | 159 | NaCl |
| K | 39.1 | 106.4 | 3.3 | 128 | $K_2O$ |
| K | 39.1 | 106.4 | 5.2 | 203 | KCl |
| $CO_3$ | 30.0 | 81.6 | — | — | — |
| $HCO_3$ | 61.0 | 165.9 | — | — | — |
| Cl | 35.5 | 96.6 | — | — | — |
| $SO_4$ | 48.0 | 130.6 | 0.9 | 43 | S |
| $SO_4$ | 48.0 | 130.6 | 4.9 | 234 | gypsum |

[1]A × B = C.

### Table B–20
### Calibration of Fertilizer Application Machinery
### (amounts of fertilizer per 100 feet of row)

| Rate per Acre | Row Width | | | | |
|---|---|---|---|---|---|
| | 18 In. | 24 In. | 30 In. | 36 In. | 48 In. |
| 250 lbs. | 14 oz. | 1 lb. | 1¼ lbs. | 1½ lbs. | 2 lbs. |
| 500 lbs. | 1¼ lbs. | 2 lbs. | 2½ lbs. | 3½ lbs. | 4½ lbs. |
| 750 lbs. | 2½ lbs. | 3 lbs. | 3¾ lbs. | 4½ lbs. | 7 lbs. |
| 1,000 lbs. | 3 lbs. | 4½ lbs. | 5¾ lbs. | 7 lbs. | 9 lbs. |
| 1,500 lbs. | 5 lbs. | 6½ lbs. | 8½ lbs. | 10½ lbs. | 14 lbs. |
| 2,000 lbs. | 6½ lbs. | 9½ lbs. | 11 lbs. | 13½ lbs. | 18 lbs. |
| 2,500 lbs. | 8½ lbs. | 11½ lbs. | 14½ lbs. | 17 lbs. | 23 lbs. |
| 3,000 lbs. | 10½ lbs. | 14 lbs. | 17½ lbs. | 21 lbs. | 28 lbs. |

Table B–21
Calibration of Liquid Fertilizer Flow[1]

| Gallons per Hour Wanted | Seconds to Fill 4-Ounce Jar | Seconds to Fill 8-Ounce Jar |
|:---:|:---:|:---:|
| ½ | 225 | 450 |
| 1 | 112 | 224 |
| 2 | 56 | 112 |
| 3 | 38 | 76 |
| 4 | 28 | 56 |
| 5 | 22 | 44 |
| 6 | 18 | 36 |
| 7 | 16 | 32 |
| 8 | 14 | 28 |
| 9 | 12 | 24 |
| 10 | 11 | 22 |
| 12 | 9 | 18 |
| 14 | 8 | 16 |
| 16 | 7 | 14 |
| 18 | 6 | 12 |
| 20 | 5.5 | 11 |

[1]To calculate the rate of application required, the following formula can be used:
Acreage × lbs. or gal. per acre + set time = lbs. or gal. per hour.

Table B–22
Calibration of Fertilizer Rate Through Sprinkler
Irrigation Systems

| Lateral Length in Feet | No. of Sprinklers (at 40-Foot Spacing) | Area Covered by 60-Foot Setting in Acres | Quantity to Apply per Setting, for Rate of 100 Pounds per Acre |
|---|---|---|---|
| 160 | 4 | 0.22 | 22 lbs. |
| 240 | 6 | 0.33 | 33 lbs. |
| 320 | 8 | 0.44 | 44 lbs. |
| 400 | 10 | 0.55 | 55 lbs. |
| 480 | 12 | 0.66 | 66 lbs. |
| 560 | 14 | 0.77 | 77 lbs. |
| 640 | 16 | 0.88 | 88 lbs. |
| 720 | 18 | 0.99 | 99 lbs. |
| 800 | 20 | 1.10 | 110 lbs. |
| 880 | 22 | 1.21 | 121 lbs. |
| 960 | 24 | 1.32 | 132 lbs. |

*Example*

To apply fertilizer at the rate of 300 lbs. per acre with 400 ft. of lateral moved at 60-ft. setting, lbs. of fertilizer applied at each setting of the lateral are calculated as follows:

Opposite a lateral length of 400 ft., find 55 lbs. of fertilzer to be applied per setting. Multiply 55 by 3 to give 165 lbs. to apply at each setting of the lateral for 300 lbs. per acre.

## Table B–23
## Estimating Amounts of Dry Bulk Fertilizer

I. Horizontal Storage

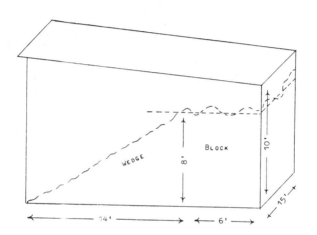

To calculate the volume in a bin:

1. Compute quantity in the block by multiplying length × width × depth. 6′ × 15′ × 8′ = 720 cu ft.
2. Compute quantity in the wedge by multiplying length × width × depth and dividing by 2. 14′ × 15′ × 8′ ÷ 2 = 840 cu ft.
3. Add the two volumes and multiply the total by the product density in lbs. per cu ft.

| | | |
|---|---|---|
| block | = | 720 cu ft. |
| wedge | = | 840 cu ft. |
| total | = | 1,560 cu ft. |

1,560 × 64 (density of ammonium sulfate) = 99,840 lbs. of product; 99,840 ÷ 2,000 = 49.92 tons.

(Continued)

### Table B–23 (Continued)

II.  Cone-shaped Pile

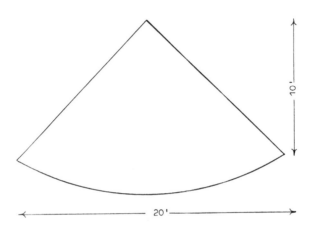

To calculate the volume in a pile:

1.  Use the mathematical formula for finding the volume of a cone:

$$\frac{\pi r^2 h}{3}$$

   Where: $\pi$ is a constant = 3.14.
   r is the radius ($\frac{1}{2}$ the diameter) of the cone measured at the base.
   h is the height of the cone.

$$\frac{3.14 \times 100 \times 10}{3} = 1,047 \text{cu ft.}$$

2.  Multiply the volume by the product density in lbs. per cu ft.
   1,047 × 66 (density of ammonium sulfate) = 69,102 lbs. of product;
   69,102 ÷ 2,000 = 34.55 tons.

Table B–24
Product Density[1]

| Product | Density |
|---------|---------|
| | *(lbs. per cu ft.)* |
| Ammonium nitrate, prilled | 52 |
| Ammonium nitrate, granulated | 62 |
| Ammonium sulfate | 64–68 |
| Calcium nitrate | 68–72 |
| Urea | 46–49 |
| Superphosphate | 68–73 |
| Treble superphosphate | 62–70 |
| 16-20-0 | 61–65 |
| 11-48-0 | 56–67 |
| 11-52-0 | 61 |
| 18-46-0 | 58–67 |
| Muriate of potash, standard | 67–75 |
| Muriate of potash, granulated | 62–68 |
| Muriate of potash, coarse | 64–69 |
| Sulfate of potash, standard | 93 |
| Sulfate of potasth, coarse | 72 |

[1]The values listed are average densities which may vary slightly, depending upon compaction of the product.

# Glossary of Terms

AAPFCO – American Association of Plant Food Control Officials.

ABSORPTION – The process by which a substance is taken into and included within another substance, e.g., intake of water by soil, or intake of gases, water, nutrients or other substances by plants.

ACID-FORMING – A term applied to any fertilizer that tends to make the soil more acid.

ACID SOIL – A soil with a pH value below 7.0. A soil having a preponderance of hydrogen over hydroxyl ions in the soil solution.

ACTIVATED SEWAGE SLUDGE – An organic fertilizer made from sewage freed from grit and coarse solids and aerated after being inoculated with microorganisms. The resulting flocculated organic matter is withdrawn from the tanks, filtered with or without the aid of coagulants, dried, ground, and screened.

ADSORPTION – The increased concentration of molecules or ions at a surface, including exchangeable cations and anions on soil particles.

AERATION, SOIL – The exchange of air in soil with air from the atmosphere. The composition of the air in a well-aerated soil is similar to that in the atmosphere; in a poorly aerated soil, the air in the soil is considerably higher in carbon dioxide and lower in oxygen than the atmosphere above the soil.

AGGREGATE – A group of soil particles cohering so as to behave mechanically as a unit.

AGRICULTURAL CHEMICAL – Synthetic or naturally occurring compounds used in agriculture to control pests, including insecticides, herbicides, biocides, bactericides, nematicides, and rodenticides.

ALKALINE – A basic reaction in which the pH reading is above 7.0, as distinguished from acidic reaction in which the pH reading is below 7.0.

ALKALINE SOIL – A soil that has an alkaline reaction, i.e., a soil for which the pH reading of the saturated soil paste is above 7.0.

ALKALI SOIL – See SODIC SOIL.

AMENDMENT – Any material, such as lime, gypsum, sawdust or synthetic conditioners, that is worked into the soil to make it more productive. Strictly, a fertilizer is also an amendment, but the term *amendment* is used more commonly for added materials other than fertilizer.

AMINO ACIDS – Nitrogen-containing organic compounds, large numbers of which link together in the formation of the protein molecule. Each amino acid molecule contains one or more amino ($-NH_2$) groups and at least one carboxyl ($-COOH$) group. In addition, some amino acids (cystine, cysteine, and methionine) contain sulfur.

AMMONIATION – A process wherein ammonia (anhydrous, aqua or a solution containing ammonia and other forms of nitrogen) is used to treat superphosphate to form ammoniated superphosphate, or to treat a mixture of fertilizer ingredients (including phosphoric acid) in the manufacture of multinutrient fertilizer.

AMMONIFICATION – Formation of ammonium compounds or ammonia.

AMMONIUM CITRATE $[(NH_4)_3C_6H_5O_7]$ – A salt formed from ammonia and citric acid. A neutral ammonium citrate solution, prepared by the official methods of the AOAC, is used as a reagent in the determination of "available" phosphate in fertilizers. After a sample is washed with water to remove the water-soluble phosphate ($P_2O_5$), the residue is treated with the neutral ammonium citrate solutions, as prescribed by the official methods, and the phosphate removed by this extraction is termed "citrate-soluble." The sum of the water-soluble plus the citrate-soluble phosphate is termed "available."

ANALYSIS – The percentage composition as found by chemical analysis, expressed in those terms that the law requires and permits. Although "analysis" and "grade" sometimes are used synonymously, the term "grade" is applied only to the three primary plant foods – nitrogen (N), phosphate ($P_2O_5$), and potash ($K_2O$) – and is stated as the guaranteed minimum quantities present. (See also GRADE.)

ANGLE OF REPOSE – The angle between the horizontal and the slope of a pile of loose material at equilibrium.

ANION – An ion carrying a negative charge of electricity.

AOAC – Association of Official Analytical Chemists (of North America).

APATITE (rock phosphate) – A mineral phosphate having the type formula $Ca_{10}(X_2)(PO_4)_6$ where X is usually fluorine, chlorine, or the hydroxyl group, either singly or together. Fluorapatite is widely distributed as the crystalline mineral and as amorphous phosphate rock, both forms of which are important fertilizer materials. Crystalline fluorapatite contains from 38.0 to 41.0 percent phosphate ($P_2O_5$) and from 3.2 to 4.3 percent fluorine. Calcium hydroxyapatite or calcium hydroxy-phosphate, $Ca_{10}(OH)_2(PO_4)_6$, may be formed to a small extent in ammoniated superphosphate.

ATTAPULGITE CLAY – A gelling clay used in suspension fertilizer.

AVAILABLE – In general, a form capable of being assimilated by a growing plant. Available nitrogen is defined as the nitrogen that is water-soluble plus what can be made soluble or converted into free ammonia. Available phosphate is that portion which is water-soluble plus the part which is soluble in ammonium citrate. Available potash is defined as that portion soluble in water or in a solution of ammonium oxalate.

AVAILABLE NUTRIENT IN SOILS — The part of the supply of a plant nutrient in the soil that can be taken up by plants at rates and in amounts significant to plant growth.

AVAILABLE WATER IN SOILS — The part of the water in the soil that can be taken up by plants at rates significant to their growth; usable; obtainable.

BASE EXCHANGE — The replacement of basic cations (Ca, Mg, Na, and K), held on the soil complex, by other basic cations. (See also CATION EXCHANGE CAPACITY.)

BASIC SLAG — A by-product in the manufacture of steel, containing lime, phosphate, and small amounts of other plant food elements such as sulfur, manganese, and iron. Basic slags may contain from 10 to 17 percent phosphate ($P_2O_5$), 35 to 50 percent calcium oxide ($CaO$) and 2 to 10 percent magnesium oxide ($MgO$). The available phosphate content of most American slag is in the range of 8 to 10 percent.

BEST MANAGEMENT PRACTICES (BMPS) — Those practices that combine scientific research with practical knowledge to optimize yields and increase crop quality while maintaining environmental integrity.

BIO-REMEDIATION — Use of living organisms to clean contaminated areas of target compounds.

BONE MEAL — Raw bone meal is cooked bones ground to a meal without any of the gelatin or glue removed. Steamed bone meal has been steamed under pressure to dissolve out part of the gelatin.

BRAND — The trade name assigned by a manufacturer to a particular fertilizer product.

BRIMSTONE – Sulfur.

BUFFER CAPACITY OF SOILS – The ability of the soil to resist a change in its pH (hydrogen ion concentration) when acid-forming or base-forming materials are added to the soil.

BULK BLEND – Physical mix of two or more dry fertilizer materials.

BULK BLENDING – The practice of mixing dry, individual, granular materials or granulated bases. The product is a mixture of granular materials rather than a granulated mixture.

BULK DENSITY – The ratio of the mass of water-free soil to its bulk volume. Bulk density is expressed in pounds per cubic foot or grams per cubic centimeter and is sometimes referred to as "apparent density." When expressed in grams per cubic centimeter, bulk density is numerically equal to apparent specific gravity or volume weight.

CALCAREOUS SOIL – A soil containing calcium carbonate, or a soil alkaline in reaction because of the presence of calcium carbonate; a soil containing enough calcium carbonate to effervesce (fizz) when treated with dilute hydrochloric acid.

CALCIUM CARBONATE EQUIVALENT – The amount of calcium carbonate required to neutralize the acidity produced by a given quantity of fertilizer product.

CARBOHYDRATE – A compound containing carbon, hydrogen, and oxygen. Usually the hydrogen and oxygen occur in the proportion of 2 to 1, such as in glucose ($C_6H_{12}O_6$).

CARBON:NITROGEN RATIO – The ratio obtained by dividing the percentage of organic carbon by the percentage of nitrogen.

CATION – An ion carrying a positive charge of electricity.

Common soil cations are calcium, magnesium, sodium, potassium, and hydrogen.

CATION EXCHANGE CAPACITY – The total quantity of cations which a soil can adsorb by cation exchange, usually expressed as milliequivalents per 100 grams.

CHELATES – Certain organic chemicals, known as chelating agents, form ring compounds in which a polyvalent metal is held between two or more atoms. Such rings are chelates. Among the best chelating agents known are ethylenediaminetetraacetic acid (EDTA), hydroxyethylenediaminetriacetic acid (HEDTA), and diethylenetriaminepentaacetic acid (DTPA).

CHLOROSIS – Yellowing of green portions of a plant, particularly the leaves.

CITRATE-SOLUBLE PHOSPHATE – That fraction of the phosphate insoluble in water but soluble in neutral ammonium citrate. However, since that soluble in water is also soluble in ammonium citrate, "citrate-soluble" may be used to indicate the sum of water-soluble plus citrate-soluble phosphate. (See also AVAILABLE.)

CLAY – A minute soil particle less than 0.002 millimeter in diameter.

COLLOID – The soil particles (inorganic or organic) having small diameters ranging from 0.20 to 0.005 micron. Colloids are characterized by high ion exchange.

CLEAR LIQUID SOLUTION – One or more plant nutrients in solution with no suspended particles.

COMPLETE FERTILIZER – A fertilizer containing all three of the primary fertilizer nutrients (nitrogen, phosphate, and potash) in sufficient amounts to be of value as nutrients.

CONDITIONER (of fertilizer) – A material added to a fertilizer to prevent caking and to keep it free-flowing.

CONDUCTIVITY, ELECTRICAL – A physical quantity that measures the readiness with which a medium transmits electricity. Commonly used for expressing the salinity of irrigation waters and soil extracts because it can be directly related to salt concentration. It is expressed in decisiemens per meter (dS/m), or in millisiemens per centimeter (mS/cm), or millimhos per centimeter (mmhos/cm), at 25°C.

CONSERVATION TILLAGE – Any system that leaves at least 30 percent of the soil surface covered with crop residue after planting.

CRITICAL NUTRIENT RANGE (CNR) – That range of concentrations above which it is reasonably certain the crop is amply supplied in a selected nutrient and below which it is reasonably certain the crop is deficient.

CURING – The process by which superphosphate or mixed fertilizers are stored until the chemical reactions have run to, or nearly to, completion.

CYTOPLASM – The portion of the protoplasm of a cell outside the nucleus.

DAMPING OFF – Sudden wilting and death of seedling plants resulting from attack by microorganisms.

DENITRIFICATION – The process by which nitrates or nitrites in the soil or organic deposits are reduced to lower oxides of nitrogen by bacterial action. The process results in the escape of nitrogen into the air.

DISTRIBUTION UNIFORMITY – Variation or non-uniformity in the amount of irrigation water applied to locations within the irrigated area. Commonly expressed numerically as the average of the lowest $\frac{1}{4}$ of the irrigation amounts divided by the average of the entire irrigation.

DOLOMITE – A material used for liming soils. Made by grinding dolomitic limestone, which contains both magnesium carbonate, $MgCO_3$, and calcium carbonate, $CaCO_3$. (See also LIME.)

ECOLOGY — The branch of biology that deals with the mutual relations among organisms and between organisms and their environment.

ELEMENTAL GUARANTEES — See GUARANTEES.

ENVIRONMENT — All external conditions that may act upon an organism or soil to influence its development, including sunlight, temperature, moisture, and other conditions.

ENZYMES — Protein substances produced by living cells which can change the rate of chemical reactions. They are organic catalysts.

EROSION — The wearing away of the land surface by detachment and transport of soil and rock materials through the action of moving water, wind, or other geological agents.

EVAPOTRANSPIRATION — The loss of water from a soil by evaporation and plant transpiration.

EXCHANGEABLE IONS — Ions held on the soil complex that may be replaced by other ions of like charge. Ions which are held so tightly that they cannot be exchanged are called nonexchangeable.

EXCHANGEABLE SODIUM PERCENTAGE — The degree of saturation of the soil exchange complex with sodium. It may be calculated by the formula:

$$ESP = \frac{Exchangeable\ sodium\,(me/100\ g\ soil)}{Cation exchange\ capacity\,(me/100\ g\ soil)} \times 100$$

FALLOW — Cropland left idle in order to restore productivity, mainly through accumulation of water, nutrients, or both. Summer fallow is a common stage before cereal grain in regions of limited rainfall. Bush or forest fallow is a rest period under woody vegetation between crops.

FERTIGATION — Application of fertilizer materials through the irrigation system.

FERTILIZER — Any natural or manufactured material added to the soil in order to supply one or more plant nutrients. The term is generally applied to inorganic materials other than lime or gypsum sold in the trade.

FERTILIZER FORMULA — The quantity and grade of materials used in making a fertilizer mixture.

FERTILIZER GRADE — An expression that indicates the weight percentage of plant nutrients in a fertilizer. Thus a 10-20-10 grade contains 10 percent nitrogen (N), 20 percent phosphate ($P_2O_5$), and 10 percent potash ($K_2O$).

FERTILIZER RATIO — The relative proportions of primary nutrients in a fertilizer grade divided by the highest common divisor for that grade; e.g., grades 10-6-4 and 20-12-8 have the ratio 5-3-2.

FIELD MOISTURE CAPACITY — The moisture content of soil in the field two or three days after a thorough wetting of the soil profile by rain or irrigation water. Field capacity is expressed as moisture percentage, dry-weight basis.

FIFTEEN-ATMOSPHERE PERCENTAGE — The moisture percentage, dry-weight basis, of a soil sample which has been wetted and brought to equilibrium in a pressure-membrane apparatus at a pressure of 221 psi. This characteristic moisture value for soils approximates the lower limit of water available for plant growth. (See also PERMANENT WILTING PERCENTAGE.)

FIXATION — The process by which available plant nutrients are rendered unavailable or "fixed" in the soil. Generally, the process by which potassium, phosphorus, and ammonium are rendered unavailable in the soil. Also, the process by which free nitrogen is chemically combined, either naturally or synthetically. (See also REVERSION and NITROGEN FIXATION.)

FLUID FERTILIZER — See liquid fertilizer and suspension fertilizer.

FORAGE — Unharvested plant material which can be used as feed by domestic animals. Forage may be grazed or cut for hay.

GRADE — The guaranteed analysis of a fertilizer containing one or more of the primary plant nutrient elements. Grades are stated in terms of the guaranteed percentages of nitrogen (N), phosphate ($P_2O_5$), and potash ($K_2O$), in that order. For example, a 10-10-10 grade would contain 10 percent nitrogen, 10 percent available phosphate, and 10 percent potash. (See also ANALYSIS.)

GROUND WATER — Water within the saturated zone of earth that supplies wells and springs and is free to move under the influence of gravity.

GUANO — The decomposed dried excrement of birds and bats, used for fertilizer purposes. It is high in nitrogen and phosphate and at one time was a major fertilizer in this country.

GUARANTEES — The AAPFCO official regulation follows: The statement of guarantees of mixed fertilizer shall be given in whole numbers. All fertilizer components with the exception of potash ($K_2O$) and phosphate ($P_2O_5$), if guaranteed, shall be stated in terms of the elements.

GYPSUM ($CaSO_4 \bullet 2H_2O$) — The common name for calcium sulfate, a mineral used in the fertilizer industry as a source of calcium and sulfur. Gypsum also is used widely in reclaiming sodic soils in the western United States. Gypsum cannot be used as a liming material, but it may reduce the alkalinity of sodic soils by replacing sodium with calcium. Another common name is landplaster. When pure it contains approximately 18.6 percent sulfur.

HARDPAN — A hardened or cemented soil horizon or layer. The soil material may be sandy or clayey and may be cemented by iron oxide, silica, calcium carbonate or other substances.

HEAVY METALS — Metallic elements in the transitional series of the periodic chart. They are not essential for plant nutrition and are usually found in small quantities in nature. They can be toxic to plants in high concentrations and to animals and humans if the concentration in the diet exceeds critical standards. Examples: cadmium, chromium, lead, nickel, and vanadium.

HOMOGENOUS DRY PRODUCT — Each granule or pellet has the same analysis.

HORIZON, SOIL — A layer of soil, approximately parallel to the soil surface, with distinct characteristics produced by soil-forming processes.

HUMUS — The well-decomposed, more or less stable portion of the organic matter in mineral soils.

HYDROGEN ION CONCENTRATION — See pH.

HYDROLYSIS — Chemical decomposition in which a compound is broken down and changed into other compounds by taking on the elements of water.

HYGROSCOPIC — Capable of taking up moisture from the air.

INORGANIC — Substances occurring as minerals in nature or obtainable from them by chemical means. Refers to all matter except the compounds of carbon, but includes carbonates.

INSOLUBLE — Not soluble. As applied to phosphate in fertilizer, that portion of the total phosphate which is soluble neither in water nor in neutral ammonium citrate. As applied to potash and nitrogen, not soluble in water.

INTEGRATED PEST MANAGEMENT — Judicious use of all available biological, physical, and chemical controls and crop rotations to reduce losses to crops caused by pests.

ION – An electrically charged particle. As used in soils, an ion refers to an electrically charged element or combination of elements resulting from the breaking up of an electrolyte in solution. Since most soil solutions are very dilute, many of the salts exist as ions. For example, all or part of the potassium chloride (muriate of potash) in most soils exists as potassium ions and chloride ions. The positively charged potassium ion is a cation, and the negatively charged chloride ion is an anion.

IRRIGATION EFFICIENCY – Percentage of the total amount of irrigation water which is beneficially used. Beneficial uses include crop utilization, leaching for salinity control, and irrigation for climate (frost) control.

KELP – Any of several species of seaweed sometimes harvested for use as a fertilizer. Dried kelp will usually contain 1.6 to 3.3 percent nitrogen, 1 to 2 percent $P_2O_5$, and 15 to 20 percent $K_2O$.

LEACHING – The removal of materials in solution by the passage of water through soil.

LEACHING REQUIREMENT – The fraction of the water entering the soil that must pass through the root zone in order to prevent soil salinity from exceeding a specified value. Leaching requirement is used primarily under steady-state or long-time average conditions.

LIME – Generally the term "lime," or "agricultural lime," is applied to ground limestone (calcium carbonate), hydrated lime (calcium hydroxide) or burned lime (calcium oxide), with or without mixtures of magnesium carbonate, magnesium hydroxide or magnesium oxide, and to materials such as basic slag, used as amendments to reduce the acidity of acid soils. In strict chemical terminology, lime refers to calcium oxide (CaO), but by an extension of meaning it is now used for all limestone-derived materials applied to neutralize acid soils.

LIME REQUIREMENT — The amount of standard ground limestone required to bring a 6.5-inch layer of an acre (about 2 million pounds in mineral soils) of acid soil to some specific lesser degree of acidity, usually to slightly or very slightly acid. In common practice, lime requirements are given in tons per acre of nearly pure limestone, ground finely enough so that all of it passes a 10-mesh screen and at least half of it passes a 100-mesh screen.

LIQUID FERTILIZER — A fluid in which the plant nutrients are in true solution.

LOAM — The textural class name for soil having a moderate amount of sand, silt, and clay. Loam soils contain 7 to 27 percent clay, 28 to 50 percent silt, and less than 52 percent sand. (In the old literature, especially English literature, the term "loam" applied to mellow soils rich in organic matter, regardless of the texture. As used in the United States, the term refers only to the relative amounts of sand, silt, and clay; loam soils may or may not be mellow.)

LUXURY CONSUMPTION — The uptake by a plant of an essential nutrient in amounts exceeding what it needs. Thus, if potassium is abundant in the soil, alfalfa may take in more than is required.

MACRONUTRIENTS — Nutrients that plants require in relatively large amounts. Essential macronutrients are nitrogen, phosphorus, and potassium.

MANURE — Generally, the refuse from stables and barnyards, including both animal excreta and straw or other litter. In some other countries the term "manure" is used more broadly and includes both farmyard or animal manure and "chemical manures," for which the term "fertilizer" is used in the United States.

MARL – An earthy deposit, consisting mainly of calcium carbonate, commonly mixed with clay or other impurities. It is formed chiefly at the margins of freshwater lakes. It is commonly used for liming acid soils.

MAXIMUM ECONOMIC YIELD (MEY) – Yield at which unit costs of production are lowered to the point of highest net return per acre . . . the most profitable yield. The MEY is achieved through implementation of best management practices.

MICRONUTRIENTS – Nutrients that plants need in only small or trace amounts. Essential micronutrients are boron, chlorine, copper, iron, manganese, molybdenum, and zinc.

MICRO-IRRIGATION – One of a number of closed irrigation systems characterized by low operating pressure (less than 40 psi), small orifice size, and constructed, in part, from plastic materials. Examples are drip, micro-sprinkler, mister, bubbler, and fogger.

MILLIEQUIVALENT or MILLIGRAM EQUIVALENT (me) – One-thousandth of an equivalent. In the case of sodium chloride, 1 me would be 0.023 gram of sodium and 0.0355 gram of chloride in 1 liter of water.

MUCK – Highly decomposed organic soil material developed from peat. Generally, muck has a higher mineral or ash content than peat and is decomposed to the point that the original plant parts cannot be identified.

NITRIFICATION – The formation of nitrites and nitrates from ammonia (or ammonium compounds), in soils by microorganisms.

NITROGEN FIXATION – Generally, the conversion of free nitrogen to nitrogen compounds. Specifically in soils, the assimilation of free nitrogen from the soil air by soil organisms and the formation of nitrogen compounds that eventually become available to plants. The nitrogen-fixing organisms associated with legumes are called symbiotic; those not definitely associated with higher plants are non-symbiotic or free-living.

NONSALINE-SODIC SOIL — A soil which contains sufficient exchangeable sodium to interfere with the growth of most crop plants, but does not contain appreciable quantities of soluble salts. The exchangeable sodium percentage is greater than 15, the conductivity of the saturation extract is less than 4 decisiemens per meter (at 25°C), and the pH of the saturated soil paste usually ranges between 8.5 and 10.0.

NUTRIENT, PLANT — Any element taken in by a plant which is essential to its growth and which is used by the plant in elaboration of its food and tissue.

ORGANIC — Compounds of carbon other than the inorganic carbonates.

ORGANIC SOIL — A general term applied to a soil or to a soil horizon that consists primarily of organic matter, such as peat soils, muck soils, and peaty soil layers.

ORTHOPHOSPHATE — A salt of orthophosphoric acid such as ammonium, calcium or potassium phosphate. Each molecule contains a single atom of phosphorus.

ORTHOPHOSPHORIC ACID — $H_3PO_4$.

PARENT MATERIAL — The unconsolidated mass of rock material (or peat) from which the soil profile develops.

PARTICLE DENSITY — The average density of the soil particles. Particle density is usually expressed in grams per cubic centimeter and is sometimes referred to as "real density" or "grain density."

PARTS PER MILLION (ppm) — A notation for indicating small amounts of materials. The expression gives the number of units by weight of the substance per million weight units of another substance, such as oven-dry soil. The term may be used to express the number of weight units of a substance per million weight units of a solution. The approximate weight of soil is 2 million pounds per acre −6 inches. Therefore, ppm × 2 equals pounds per acre −6 inches of soil, or ppm × 4 equals pounds per acre-foot of soil.

PEAT — Peat is partly decayed vegetable matter of natural occurrence. It is composed chiefly of organic matter that contains some nitrogen of low activity. (AAPFCO).

PERCOLATION — The downward movement of water through soil.

PERMANENT WILTING PERCENTAGE — The moisture percentage of soil at which plants wilt and fail to recover turgidity (15 atmospheres). The expression has significance only for non-saline soils.

PERMEABILITY, SOIL — The quality of a soil horizon that enables water or air to move through it. It can be measured quantitatively in terms of rate of flow of water through a unit cross section in unit time under specified temperature and hydraulic conditions. Values for saturated soils usually are called hydraulic conductivity. The permeability of a soil is controlled by the least permeable horizon.

pH — A numerical designation of acidity and alkalinity. Technically, pH is the common logarithm of the reciprocal of the hydrogen ion concentration of a solution. A pH of 7.0 indicates precise neutrality; higher values indicate increasing alkalinity, and lower values indicate increasing acidity.

PHOSPHATE — A salt of phosphoric acid made by combining phosphoric acid with ions such as ammonium, calcium, potassium or sodium.

PHOSPHATE ROCK — Phosphate-bearing ore composed largely of tricalcium phosphate. Phosphate rock can be treated with strong acids or heat to make available forms of phosphate.

Finely ground rock phosphate is sometimes used in long-term fertility programs.

PHOTOSYNTHESIS – The process by which green plants combine water and carbon dioxide to form carbohydrates under the action of light. Chlorophyll is required for the conversion of light energy into chemical energy.

POLYPHOSPHATE – A salt of polyphosphoric acid such as ammonium, calcium or potassium polyphosphate. *Poly* means "many" and refers to multiple linkages of phosphorus in each molecule.

POLYPHOSPHORIC ACID – Condensed phosphoric acid ranging in $P_2O_5$ content from 68 to 83 percent.

POP UP – Application of low rates of fertilizer materials at planting in direct contact with the seed to encourage early rapid growth.

POROSITY – The fraction of soil volume not occupied by soil particles.

POTASH – The AAPFCO has adopted the term potash to designate potassium oxide ($K_2O$).

PRIMARY PLANT NUTRIENTS (plant foods) – Nitrogen (N), phosphate ($P_2O_5$), and potash ($K_2O$).

PRODUCTIVITY – In simplest terms, the ability of the soil to produce. It differs from fertility to the extent that a soil may be fertile and yet unable to produce because of other limiting factors.

PROFILE, SOIL – A vertical section of the soil extending through all its horizons and into the parent material.

PROTEIN – Any of a group of high-molecular-weight, nitrogen-containing compounds that yield amino acids on hydrolysis. Protein is a vital part of living matter and is one of the essential food substances of animals.

PROTOPLASM – The jellylike substance in plant and animal cells; it is basic to all life processes.

PUDDLED SOIL – Dense, massive soil artificially compacted when wet and having no regular structure. The condition commonly results from the tillage of a clayey soil when it is wet.

PYRITE ($FeS_2$) – A mineral composed principally of iron and sulfur, with varying small amounts of other metals. "Fool's gold."

QUICK TESTS – Simple and rapid chemical tests of soils designed to give an approximation of the nutrients available to plants.

RATIO – See FERTILIZER RATIO.

RECLAMATION – The process of restoring lands to productivity by removing excess soluble salts or excess exchangeable sodium from soils.

REVERSION – The interaction of a plant nutrient with the soil which causes the nutrient to become less available. In fertilizer manufacturing, the excessive use of ammonia in ammoniation of phosphates results in phosphate reversion. (See also FIXATION.)

SALINE-SODIC SOIL – A soil containing sufficient exchangeable sodium to interfere with the growth of most crop plants and containing appreciable quantities of soluble salts. The exchangeable sodium percentage is greater than 15, and the electrical conductivity of the saturation extract is greater than 4 decisiemens per meter (at 25°C). The pH reading of the saturated soil paste is usually less than 8.5.

SALINE SOIL – A soil containing enough soluble salts to impair its productivity for plants, but not containing an excess of exchangeable sodium.

SALT INDEX – An index used to compare solubilities of chemical compounds. Most nitrogen and potash compounds have high indexes, and phosphate compounds have low indexes. When applied too close to seed or on foliage, the compounds with high indexes cause plants to wilt or die.

SALTING OUT TEMPERATURE – Temperature at which salts precipitate.

SALTS – The products, other than water, of the reaction of an acid with a base. Salts commonly found in soils break up into cations (sodium, calcium, etc.) and anions (chloride, sulfate, etc.) when dissolved in water.

SAND – Individual rock or mineral fragments in soils having diameters ranging from 0.05 millimeter to 2.0 millimeters. Usually sand grains consist chiefly of quartz, but they may be of any mineral composition. The textural class name of any soil that contains 85 percent or more sand and not more than 10 percent clay.

SATURATED SOIL PASTE – A particular mixture of soil and water commonly used for measurements and for obtaining soil extracts. At saturation, the soil paste glistens as it reflects light, flows slightly when the container is tipped, and slides freely and cleanly from a spatula for all soils, except those with high clay content.

SATURATION EXTRACT – The solution extracted from a soil at its saturation percentage.

SATURATION PERCENTAGE – The moisture percentage of a saturated soil paste, expressed on a dry-weight basis.

SECONDARY PLANT NUTRIENTS – Calcium, magnesium, and sulfur.

SEGREGATION – Separation of one component or raw material from another, such as in a dry bulk blend.

SEPARATE, SOIL – One of the individual-size groups of mineral soil particles – sand, silt or clay.

SERIES, SOIL — A group of soils that have soil horizons similar in their differentiating characteristics and arrangement in the soil profile, except for the texture of the surface soil, and are formed from a particular type of parent material. Soil series is an important category in detailed soil classification. Individual series are given proper names from places near the first recorded occurrence. Thus, names like Yolo, Panoche, Hanford, and San Joaquin are names of soil series that appear on soil maps, and each connotes a unique combination of many soil characteristics.

SEWAGE SLUDGE — An organic product resulting from the treatment of sewage. The composition varies widely depending on the method of treatment.

SILT — (1) Individual mineral particles of soil that range in diameter between the upper size of clay, 0.002 mm, and the lower size of very fine sand, 0.05 mm. (2) Soil of the textural class silt containing 80 percent or more silt and less than 12 percent clay. (3) Sediments deposited from water in which the individual grains are approximately the size of silt, although the term is sometimes applied loosely to sediments containing considerable sand and clay.

SLURRY FERTILIZER — A fluid mixture containing dissolved and undissolved plant nutrient materials which requires continuous mechanical agitation to assure homogeneity.

SODIC SOIL — A soil that contains sufficient exchangeable sodium to interfere with the growth of most plants, either with or without appreciable quantities of soluble salts. (See also NONSALINE-SODIC SOIL and SALINE-SODIC SOIL.)

SODIUM ADSORPTION RATIO — A ratio for soil extracts and irrigation waters used to express the relative activity of sodium ions in exchange reactions with soil.

$$SAR = \frac{Na^+}{\sqrt{(Ca^{++} + Mg^{++})/2}}$$

The ionic concentrations are expressed in milliequivalents per liter.

SODIUM PERCENTAGE – The percent sodium of total cations. Calculations are based on milliequivalents rather than weight.

SOIL MOISTURE STRESS – The sum of the soil moisture tension and the osmotic pressure of the soil solution. It is the force plants must overcome to withdraw moisture from the soil.

SOIL MOISTURE TENSION – The force by which moisture is held in the soil. It is a negative pressure and may be expressed in any convenient pressure unit. Tension does not include osmotic pressure values.

STRUCTURE, SOIL – The physical arrangement of the soil particles.

SUBSOIL – Roughly, that part of the soil below plow depth.

SUPERPHOSPHATE – Superphosphate is a product obtained by mixing rock phosphate with either sulfuric acid or phosphoric acid or with both acids. The grade that shows the available phosphate shall be used as a prefix to the name. Example: 20 percent superphosphate. (AAPFCO)

SUPERPHOSPHORIC ACID – See POLYPHOSPHORIC ACID.

SURFACE WATER – Water on or above the ground including rivers, lakes, canals, and reservoirs.

SUSPENSION FERTILIZER – A fluid containing dissolved and undissolved plant nutrients. The suspension of the undissolved plant nutrients may be inherent to the materials or produced with the aid of a suspending agent of non-fertilizer properties. Mechanical agitation may be necessary in some cases to facilitate uniform suspension of undissolved plant nutrients.

SUSTAINABLE AGRICULTURE – The production of crops with

judicious use of all inputs to maintain production indefinitely.

SYMBIOSIS — The living together of two different organisms with a resulting mutual benefit. A common example is the association of rhizobia with legumes; the resulting nitrogen fixation is sometimes called symbiotic nitrogen fixation. Adjective: symbiotic.

TANKAGE — Dried animal residue. Process tankage is made from leather scrap, wool, and other inert nitrogenous materials by steaming under pressure with or without addition of acid. This treatment increases the availability of the nitrogen to plants.

TENSIOMETER — A device used to measure the tension with which water is held in the soil.

TEXTURE, SOIL — The relative proportions of the various size groups of individual soil grains in a mass of soil. Specifically, it refers to the proportions of sand, silt, and clay.

TILTH — The physical condition of a soil with respect to its fitness for the growth of plants.

TRACE ELEMENTS — See MICRONUTRIENTS.

TRANSPIRATION — Loss of water vapor from the leaves and stems of living plants to the atmosphere.

UNIT — The AOAC has adopted as official the following definition: A unit of plant food is twenty (20) pounds, or one percent (1 percent) of a ton.

VAPOR PRESSURE — The pressure exerted above a liquid because of the tendency of vapor to escape from the surface. Typical examples are the pressure above liquid anhydrous ammonia or ammonia-ammonium nitrate solutions. A negative value indicates that the vapor pressure above the liquid is less than atmospheric pressure. Vapor pressure is temperature-dependent. Increasing the temperature increases the vapor pressure above the liquid.

VOLATILIZATION — The evaporation or changing of a substance from liquid to vapor.

WATER TABLE — The upper surface of ground water.

WATER TABLE, PERCHED — The upper surface of a body of free ground water in a zone of saturation separated from underlying ground water by unsaturated material.

WEATHERING — The physical and chemical disintegration and decomposition of parent materials as in soil formation.

WILTING PERCENTAGE — See PERMANENT WILTING PERCENTAGE.

# Index

Absorption, 74, 305
Acid-forming amendments, 225–26, 305
Acidity
  active, 217
  plant tolerance for, 216
  potential, 217
Acid soils, 10, 216–17, 305
  lime requirement for, 217–18
  liming materials for, 218–19
  pH range for, 217
Acre, number of trees or plants on, 290
Active acidity, 217
ADP-ATP reaction, 77
Adsorption, 12, 305
Aeolian material, 5
Aeration, soil, 9, 305
Aggregate, 8-9, 15, 305
Agricultural chemical, 244, 305
Agriculture system, U.S. Department of, for classifying soil separates, 6
A horizon, 3, 4, 5
Air sparging, 156
Alfalfa
  boron deficiency on, Plate 4–15
  potassium deficiency on, Plate 4–6
  tissue analysis of, 201
Algae, 17
Alkaline soils, 10, 215, 306
Alkali soils, 10, 222, 306. See also Sodic (alkali) soil
Alluvial material, 5
Aluminum sulfate, amendment values of, 225
Amendments, 306. See also Soil amendments
  acid-forming, 225-26, 305
  chemical, 292

Aminization, 90
Amino acids, 90, 306
Ammionic nitrogen, 89
Ammonia, conversions of, to fertilizers, 110–11
Ammoniation, 126, 146–47, 306
Ammonic-N, analysis methods for, 193
Ammonification, 90, 306
Ammonium, 13
  properties of, 146
Ammonium bisulfite, 158
Ammonium carbonate, 116
Ammonium nitrate, 112, 113–14, 144
  properties of, 114
  storage and handling of, 151–52, 154
Ammonium nitrate–lime, 114
Ammonium phosphates, 126
Ammonium phosphate sulfate, 126
Ammonium polyphosphate, 150
Ammonium polysulfide, 158, 172
  amendment values of, 225
Ammonium sulfate, 114–15, 144
  properties of, 115
  storage and handling of, 152
Ammonium thiosulfate, 119, 158
AN-20, 117, 118
Analysis, 307. See also Soil testing
Anhydrous ammonia, 110–12, 167, 174, 177–78
  properties of, 111
  reaction with sulfuric acid, 114

  in side dressing applications, 170
  storage and handling of, 154
Animal manures, 16, 243, 317
  as source of organic matter, 16
  utilization of, in crop production, 244
Anions, 30, 31, 32, 307
  impact of, on soil, 35–37
Apatite. See Phosphate rock
Apple, boron deficiency on, Plate 4–16
Aqua ammonia, 112, 167, 174
  in side dressing applications, 170
  storage and handling of, 154
Ash, as liming material, 218
Asparagus
  plant food utilization by, 97–98
  tissue analysis of, 201
Atmosphere, and plant growth, 83

Bacteria
  free-living, 18
  heterotrophic, 18
  nitrogen-fixing, 18
  in soils, 17–18
Barley
  effect of salt accumulation on, 221
  plant food utilization by, 97
  tissue analysis of, 201
Basic slag, 308
B horizon, 3, 4-5
Beans
  plant food utilization by, 97–98